Lecture Notes in Bioinformatics 10849

Subseries of Lecture Notes in Computer Science

More information about this series at http://www.springer.com/series/5381

Jesper Jansson · Carlos Martín-Vide
Miguel A. Vega-Rodríguez (Eds.)

Algorithms for Computational Biology

5th International Conference, AlCoB 2018
Hong Kong, China, June 25–26, 2018
Proceedings

 Springer

Editors
Jesper Jansson ⓘ
The Hong Kong Polytechnic University
Kowloon
Hong Kong SAR, China

Miguel A. Vega-Rodríguez ⓘ
University of Extremadura
Cáceres
Spain

Carlos Martín-Vide ⓘ
Rovira i Virgili University
Tarragona
Spain

ISSN 0302-9743   ISSN 1611-3349 (electronic)
Lecture Notes in Bioinformatics
ISBN 978-3-319-91937-9   ISBN 978-3-319-91938-6 (eBook)
https://doi.org/10.1007/978-3-319-91938-6

Library of Congress Control Number: 2018943444

LNCS Sublibrary: SL8 – Bioinformatics

Printed on acid-free paper

This Springer imprint is published by the registered company Springer International Publishing AG
part of Springer Nature
The registered company address is: Gewerbestrasse 11, 6330 Cham, Switzerland

Preface

These proceedings contain the papers that were presented at the 5th International Conference on Algorithms for Computational Biology (AlCoB 2018), held in Hong Kong, during June 25–26, 2018.

The scope of AlCoB includes topics of either theoretical or applied interest, namely:

- Exact sequence analysis
- Approximate sequence analysis
- Pairwise sequence alignment
- Multiple sequence alignment
- Sequence assembly
- Genome rearrangement
- Regulatory motif finding
- Phylogeny reconstruction
- Phylogeny comparison
- Structure prediction
- Compressive genomics
- Proteomics: molecular pathways, interaction networks, mass spectrometry analysis
- Transcriptomics: splicing variants, isoform inference and quantification, differential analysis
- Next-generation sequencing: population genomics, metagenomics, metatranscriptomics
- Microbiome analysis
- Systems biology

AlCoB 2018 received 20 submissions. Most papers were reviewed by three Program Committee members. There were also a few external reviewers consulted. After a thorough and vivid discussion phase, the committee decided to accept 11 papers (which represents an acceptance rate of about 55%). The conference program included three invited talks and some poster presentations of work in progress.

The excellent facilities provided by the EasyChair conference management system allowed us to deal with the submissions successfully and handle the preparation of these proceedings in time.

We would like to thank all invited speakers and authors for their contributions, the Program Committee and the external reviewers for their cooperation, and Springer for its very professional publishing work.

April 2018

Jesper Jansson
Carlos Martín-Vide
Miguel A. Vega-Rodríguez

Organization

AlCoB 2018 was organized by the Department of Computing, Hong Kong Polytechnic University, Hong Kong, and the Research Group on Mathematical Linguistics (GRLMC), Rovira i Virgili University, Tarragona, Spain.

Program Committee

Josep Francesc Abril	University of Barcelona, Spain
Kees Albers	Radboud University Medical Centre, The Netherlands
Emmanuel Barillot	Curie Institute, France
Philipp Bucher	Swiss Federal Institute of Technology, Switzerland
Rita Casadio	University of Bologna, Italy
José C. Clemente	Icahn School of Medicine at Mount Sinai, USA
Eytan Domany	Weizmann Institute of Science, Israel
Liliana Florea	Johns Hopkins University, USA
Dmitrij Frishman	Technical University of Munich, Germany
Terry Furey	University of North Carolina at Chapel Hill, USA
Osamu Gotoh	Institute of Advanced Industrial Science and Technology, Japan
John Hancock	ELIXIR Europe, UK
Robert Harrison	Georgia State University, USA
Martijn Huynen	Radboud University Medical Centre, The Netherlands
Jesper Jansson	The Hong Kong Polytechnic University, Hong Kong, SAR China
Pouya Kheradpour	Verily Life Sciences, USA
Julien Lagarde	Centre for Genomic Regulation, Spain
Gerton Lunter	University of Oxford, UK
Ruibang Luo	Johns Hopkins University, USA
Bill Majoros	Duke University, USA
Carlos Martín-Vide (Chair)	Rovira i Virgili University, Spain
Huaiyu Mi	University of Southern California, USA
Ryan E. Mills	University of Michigan, USA
Kenta Nakai	University of Tokyo, Japan
William Stafford Noble	University of Washington, USA
Sandra Orchard	European Bioinformatics Institute, UK
Mihaela Pertea	Johns Hopkins University, USA
Itsik Pe'er	Columbia University, USA
Paolo Ribeca	The Pirbright Institute, UK
Peter Robinson	The Jackson Laboratory for Genomic Medicine, USA
Stephane Rombauts	Ghent University, Belgium
Russell Schwartz	Carnegie Mellon University, USA

Xinghua Shi	University of North Carolina at Charlotte, USA
Denis Shields	University College Dublin, Ireland
Ilya Shmulevich	Institute for Systems Biology, USA
Thomas Sicheritz-Pontén	Technical University of Denmark, Denmark
Steven Skiena	Stony Brook University, USA
Wing-Kin Sung	National University of Singapore, Singapore
Weili Wu	The University of Texas at Dallas, USA
Wenzhong Xiao	Massachusetts General Hospital, USA
Shibu Yooseph	University of Central Florida, USA
Yaoqi Zhou	Griffith University, Australia

Additional Reviewers

Bhaskar Dasgupta
Sam Kovaka
Yang Lu

Jacob Schreiber
Krister Swenson
Weichen Zhou

Organizing Committee

Jesper Jansson (Co-chair), Hong Kong
Konstantinos Mampentzidis, Hong Kong
Carlos Martín-Vide (Co-chair), Tarragona
Manuel J. Parra-Royón, Granada
David Silva, London
Sandhya T. P., Hong Kong
Miguel A. Vega-Rodríguez, Cáceres

Contents

Systems Biology and Other Biological Processes

Invited Talk

Algorithms for Analysis and Control
of Boolean Networks

Tatsuya Akutsu[✉][iD]

Bioinformatics Center, Institute for Chemical Research,
Kyoto University, Gokasho, Uji, Kyoto 611-0011, Japan
takutsu@kuicr.kyoto-u.ac.jp

Abstract. Boolean network is a discrete mathematical model of gene regulatory networks. In this short article, we briefly review algorithmic results on finding attractors in Boolean networks. Since it is known that the problem of finding a singleton attractor is NP-hard and the problem can be trivially solved in $O^*(2^n)$ time (under a reasonable assumption), we focus on special cases in which the problem can be solved in $O((2-\delta)^n)$ time for some constant $\delta > 0$. We also briefly review algorithmic results on control of Boolean networks.

Keywords: Boolean networks · Attractors · Controllability

1   Boolean Networks

Mathematical analysis of biological networks is an important topic in bioinformatics and computational biology. For that purpose, various kinds of mathematical models have been proposed. Among them, the *Boolean network* (BN) has been extensively studied since 1960's [3]. BN is a discrete mathematical model of gene regulatory networks, in which each node (e.g., gene) takes either 0 or 1 and the states of nodes change synchronously according to regulation rules given as Boolean functions, where 1 (resp., 0) means that the corresponding gene is expressed (resp., not expressed).

Formally, a BN $N(V, F)$ consists of a set $V = \{x_1, \ldots, x_n\}$ of nodes and a list $F = (f_1, \ldots, f_n)$ of *Boolean functions*, where a Boolean function $f_i(x_{i_1}, \ldots, x_{i_{k_i}})$ with inputs from specified nodes $x_{i_1}, \ldots, x_{i_{k_i}}$ is assigned to each node x_i. We use $IN(x_i)$ to denote the set of input nodes $x_{i_1}, \ldots, x_{i_k}$ to x_i. Each node takes either 0 or 1 at each discrete time t, and the state of node x_i at time t is denoted by $x_i(t)$. Then, the state of node x_i at time $t + 1$ is determined by

$$x_i(t+1) = f_i(x_{i_1}(t), \ldots, x_{i_{k_i}}(t)).$$

The state of the whole network at time step t is represented by an n-dimensional 0–1 vector $\mathbf{x}(t) = [x_1(t), \ldots, x_n(t)]$. We also write $x_i(t+1) = f_i(\mathbf{x}(t))$ to denote

Partially supported by JSPS, Japan: Grant-in-Aid 26240034.

the regulation rule for x_i and $\mathbf{x}(t + 1) = \mathbf{f}(\mathbf{x}(t))$ to denote the regulation rule for the whole BN. The network structure of a BN $N(V, F)$ is represented by a directed graph $G(V, E)$ such that $E = \{(x_{i_j}, x_i) | x_{i_j} \in IN(x_i)\}$. The dynamics of a BN can be well represented by a *state transition diagram*, in which a vertex and a directed edge correspond to a (global) state of the BN and a state transition, respectively. For example, consider a BN $N(V, F)$ defined by

$$x_1(t + 1) = x_3(t),$$
$$x_2(t + 1) = x_1(t) \wedge \overline{x_3(t)},$$
$$x_3(t + 1) = x_1(t) \wedge \overline{x_2(t)}.$$

Then, $G(V, E)$ and its state transition diagram are as in Fig. 1(A) and (B), respectively.

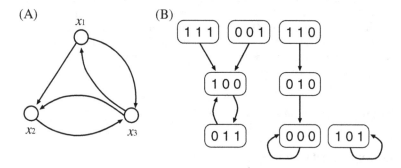

Fig. 1. Example of BN. (A) $G(V, E)$. (B) State transition diagram.

Starting from any initial state, a BN will eventually reach a cyclic sequence of states, called an *attractor*, which is often regarded as a type of a cell. An attractor consisting of only one global state (i.e., $\mathbf{x} = \mathbf{f}(\mathbf{x})$) is called a *singleton attractor*. Otherwise, it is called a *periodic attractor*. A periodic attractor consisting of p states is called a p-periodic attractor. For example, the BN given in Fig. 1 has two singleton attractors $\langle [0, 0, 0] \rangle$ and $\langle [1, 0, 1] \rangle$, and one 2-periodic attractor $\langle [0, 1, 1], [1, 0, 0] \rangle$.

2   Attractor Detection

After making a BN model of some organism or its part, it is important to find or enumerate attractors because they are considered to correspond to cell types. Since an attractor corresponds to a directed cycle in a state transition diagram, attractors can be enumerated by enumerating cycles in the diagram. Since a state transition diagram consists of 2^n vertices and 2^n edges, enumeration of attractors can be done in $O^*(2^n)$ time[1] once the diagram is constructed. Construction of

---

[1] $O^*(f(n))$ denotes $O(f(n)poly(n))$.

the diagram can also be done in $O^*(2^n)$ time if the value of each Boolean function appearing in the BN can be computed in polynomial time. On the other hand, it is known that deciding whether or not there exists a singleton attractor is NP-hard. Since it is quite difficult to break the $O^*(2^n)$ barrier for the general case, existing studies focused on developments of $O^*((2 - \delta)^n)$ time algorithms for special but important subclasses of BNs. In particular, the author and his colleagues developed the following algorithms [1]:

- $O(1.587^n)$ time algorithm for detection of a singleton attractor for a BN consisting of AND/OR functions (i.e., each Boolean function is a conjunction or disjunction of literals),
- $O(1.871^n)$ time algorithm for detection of a singleton attractor for a BN consisting of nested canalyzing functions (see [1] for the definition of a nested canalyzing function),
- $O(1.985^n)$ time algorithm for detection of a 2-periodic attractor for a BN consisting of AND/OR functions,
- $O^*(n^{2p(w+1)})$ time algorithm for detection of a p-periodic attractor for a BN consisting of nested canalyzing functions and having bounded treewidth w.

Improvements of these algorithms and developments of $O^*((2 - \delta)^n)$ time algorithms for other important subclasses are left as open problems.

3   Control of Boolean Networks

Recently, control of BNs has captured a lot of attentions because of its potential application to control of cells and diseases. In particular, algebraic approaches based on *semi-tensor product* have been extensively studied [2,4].

Here, we focus on computational complexity of control of BNs. Although there exist many variants, one of the fundamental control problems is defined as: given a BN with control nodes, its initial and target states, find a sequence of 0–1 vectors for control nodes which leads BN from the initial state to the target state. Formally, this problem is defined as follows. Let $V = \{x_1, \ldots, x_n, x_{n+1}, \ldots, x_{n+m}\}$, where $x_1, \ldots, x_n$ are internal nodes and $x_{n+1}, \ldots, x_{n+m}$ are external nodes (i.e., control nodes). We use u_i to denote an external node x_{n+i}. Let $\mathbf{x}(t) = [x_1(t), \ldots, x_n(t)]$ and $\mathbf{u}(t) = [u_1(t), \ldots, u_m(t)]$. Then, the state of each internal node $x_i(t + 1)$ $(i = 1, \ldots, n)$ at time step $t + 1$ is determined by

$$x_i(t + 1) = f_i(x_{i_1}(t), \ldots, x_{i_{k_i}}(t)),$$

where each x_{i_j} is either an internal node or an external node. We can describe the dynamics of a BN with external nodes by

$$\mathbf{x}(t + 1) = \mathbf{f}(\mathbf{x}(t), \mathbf{u}(t)).$$

Then, control of BN is defined as follows (see also Fig. 2): given a BN with external nodes, initial and target states $\mathbf{x}^0$ and $\mathbf{x}^M$ of the internal nodes, find a sequence of 0–1 vectors $\langle \mathbf{u}(0), \ldots, \mathbf{u}(M - 1) \rangle$ such that $\mathbf{x}(0) = \mathbf{x}^0$ and $\mathbf{x}(M) = \mathbf{x}^M$

(if there does not exist such a sequence, "None" should be the output). For example, consider a BN defined by

$$x_1(t+1) = \overline{u_1(t)},$$
$$x_2(t+1) = x_1(t) \wedge u_2(t),$$
$$x_3(t+1) = x_1(t) \vee x_2(t).$$

Suppose that $\mathbf{x}^0 = [1, 0, 0]$, $\mathbf{x}^M = [0.1.1]$, and $M = 3$ (see also Fig. 2). Then, we have a control sequence $\langle \mathbf{u}(0), \mathbf{u}(1), \mathbf{u}(2) \rangle$ with

$$\mathbf{u}(0) = [0, 0], \quad \mathbf{u}(1) = [0, 1], \quad \mathbf{u}(2) = [1, 1],$$

which drives the BN from $\mathbf{x}^0$ to $\mathbf{x}^M$.

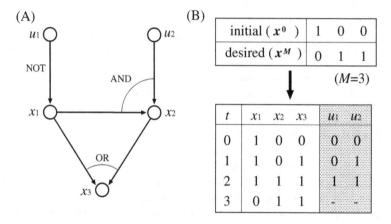

Fig. 2. Example of control of BN. (A) BN with external nodes u_1 and u_2. (B) State transitions from the initial state to the target state.

This control problem can be solved by a dynamic programming algorithm as follows. We use a table $D[b_1, \ldots, b_n, t]$, where each entry (except t) takes either 0 or 1, and t corresponds to a time step. $D[b_1, \ldots, b_n, t]$ takes 1 if there exists a control sequence $\langle \mathbf{u}(t), \ldots, \mathbf{u}(M-1) \rangle$ which leads to the target state $\mathbf{x}^M$ beginning from the state $[b_1, \ldots, b_n]$ at time t. This table is computed from $t = M$ to $t = 0$ by using the following procedure:

$$D[b_1, \ldots, b_n, M] = \begin{cases} 1, & \text{if } [b_1, \ldots, b_n] = \mathbf{x}^M, \\ 0, & \text{otherwise,} \end{cases}$$

$$D[b_1, \ldots, b_n, t-1] = \begin{cases} 1, & \text{if there exists } (\mathbf{c}, \mathbf{u}) \text{ such that } D[c_1, \ldots, c_n, t] = 1 \\ & \quad \text{and } \mathbf{c} = \mathbf{f}(\mathbf{b}, \mathbf{u}), \\ 0, & \text{otherwise,} \end{cases}$$

where $\mathbf{b} = [b_1, \ldots, b_n]$ and $\mathbf{c} = [c_1, \ldots, c_n]$. Then, there exists a desired control sequence if and only if $D[a_1, \ldots, a_n, 0] = 1$ holds for $\mathbf{x}^0 = [a_1, \ldots, a_n]$. Once the table is constructed, a desired control sequence can be obtained using the standard traceback technique. It is easy to see that this algorithm requires $O(M \cdot 2^{n+m})$ time excluding the time for calculation of Boolean functions. This is an exponential-time algorithm. Actually, the problem is NP-hard even for $M = 1$ and BNs with very simple network structures. Furthermore, it is PSPACE-hard if M is not bounded [1]. A polynomial time algorithm is known only for BNs with tree structures [1]. Development of an algorithm that is faster than the $O(M \cdot 2^{n+m})$ time one is left as an open problem, even for considerably restricted BNs.

References

1. Akutsu, T.: Algorithms for Analysis, Inference, and Control of Boolean Networks. World Scientific, Singapore (2018)
2. Cheng, D., Qi, H., Li, Z.: Analysis and Control of Boolean Networks: A Semi-tensor Product Approach. Springer, London (2011). https://doi.org/10.1007/978-0-85729-097-7
3. Kauffman, S.A.: Homeostasis and differentiation in random genetic control networks. Nature **224**, 177–178 (1969)
4. Lu, J., Li, H., Liu, Y., Li, F.: Survey on semi-tensor product method with its applications in logical networks and other finite-valued systems. IET Control Theory Appl. **11**, 2040–2047 (2017)

Phylogenetics

Reconciliation Feasibility of Non-binary Gene Trees Under a Duplication-Loss-Coalescence Model

Ricson Cheng[1], Matthew Dohlen[2], Chen Pekker[3], Gabriel Quiroz[3],
Jincheng Wang[3], Ran Libeskind-Hadas[3], and Yi-Chieh Wu[3(✉)]

[1] Department of Computer Science, Carnegie Mellon University,
5000 Forbes Ave, Pittsburgh, PA 15213, USA
`ricsonc@andrew.cmu.edu`
[2] Department of Computer Science, California Polytechnic University,
3801 W Temple Ave, Pomona, CA 91768, USA
`mdohlen@cs.hmc.edu`
[3] Department of Computer Science, Harvey Mudd College,
301 Platt Blvd, Claremont, CA 91711, USA
{`gpekker,gquiroz,jiwang`}`@hmc.edu`, {`hadas,yjw`}`@cs.hmc.edu`

Abstract. Phylogenetic tree reconciliation is a widely-used method to understand gene family evolution. For eukaryotes, the duplication-loss-coalescence (DLC) model seeks to explain incongruence between gene trees and species trees by postulating gene duplication, gene loss, and deep coalescence events. While efficient algorithms exist for inferring optimal DLC reconciliations, they assume that only one individual is sampled per species. In recent work, we demonstrated that with additional samples, there exist gene tree topologies that are impossible to reconcile with any species tree. However, our algorithm required the gene tree to be binary whereas, in practice, gene trees are often non-binary due to uncertainty in the reconstruction process. In this work, we consider for the first time reconciliation under the DLC model with non-binary gene trees. Specifically, we describe an efficient algorithm that takes as input an arbitrary gene tree with an arbitrary number of samples per species and either (1) determines that there is a valid reconcilable binary resolution of that tree and constructs one such resolution or (2) determines that there exists no valid reconcilable binary resolution of that tree. Our work makes it possible to systematically analyze non-binary gene trees and will help biologists identify incorrect gene tree topologies and thus avoid incorrect evolutionary inferences.

Keywords: Phylogenetics · Reconciliation
Gene duplication and loss · Coalescence · Non-binary trees

M. Dohlen and C. Pekker—These authors contributed equally to this work.

© Springer International Publishing AG, part of Springer Nature 2018
J. Jansson et al. (Eds.): AlCoB 2018, LNBI 10849, pp. 11–23, 2018.
https://doi.org/10.1007/978-3-319-91938-6_2

1   Introduction

Phylogenetic tree reconciliation is a fundamental technique for understanding the evolutionary histories of genes found across a set of species. Given a gene tree, species tree, and the association between their leaves, a *reconciliation* postulates evolutionary events to explain the incongruence, or topological differences, between those trees. These events may include gene duplication [21], gene loss [2], horizontal gene transfer [20], and incomplete lineage sorting [11], among others. Accurate reconciliations can provide important insights into centrally important questions on gene evolution and the introduction of new gene functions [16,30].

Reconciliations rely on an underlying evolutionary model. Some widely-used models include the *duplication-loss* (DL) model [1,6,7,14,15,22,24,35], which allows for gene duplication and gene loss; the *duplication-transfer-loss* (DTL) model [3,8–10,13,29], which considers horizontal gene transfers as well; and the *multispecies coalescent* (MSC) model [19,23,31,33], which allows for incomplete lineage sorting through deep coalescence. However, the DL and DTL models cannot address population effects, and MSC models cannot address paralogous gene families. Thus, each model has limited accuracy and applicability.

Recently, a unified *duplication-loss-coalescence* (DLC) model was proposed that combines the DL and MSC models [25], thereby addressing the most common events in eukaryotic gene evolution. Given a single haploid sample per species, two algorithms exist for solving the DLC reconciliation problem: DLCoalRecon finds the reconciliation with highest posterior probability [25], and DLCpar finds a most parsimonious reconciliation (one that minimizes the total cost of the constituent events) [32]. More recently, we extended the DLC model to allow for multiple samples per species and demonstrated that these multiple samples impose additional constraints such that gene trees may have no feasible reconciliation. Such infeasible gene trees can occur, for example, due to noisy sequencing, reconstruction error, or violations of model assumptions. To address this problem, we presented a polynomial-time algorithm for determining reconciliation feasibility of gene trees under the DLC model [26].

A significant limitation of these formulations is that they require the gene and species trees to be binary. In practice, species trees for several clades are binary since their reconstruction can benefit from well-behaved gene families as well as multigene phylogeny construction methods [4,12]. When a species tree is non-binary, the non-binary nodes, or *polytomies*, are often "hard" and represent the simultaneous speciation of a common ancestor into multiple species. In contrast, gene trees are often non-binary due to lack of phylogenetic signal [27]. Their *polytomies* are "soft" in the sense that better data would allow us to resolve such nodes to yield a binary gene tree. Note that the number of binary resolutions is exponential in the number of non-binary nodes and their maximum out-degree. When given a non-binary gene tree and a binary species tree, reconciliation algorithms under the simpler DL and DTL models often seek to find a binary resolution of the gene tree that minimizes the reconciliation cost [5,17,18,34].

In this work, we consider the problem of binary resolution under the DLC model with multiple samples per species. We present an efficient new algorithm that finds a valid binary resolution when such a resolution exists. Note that a brute-force approach of enumerating each binary resolution and testing it for reconcilability would take exponential time and thus be impractical. Using our algorithm, we also prove that there exist non-binary gene trees for which there is no valid binary resolution. This work generalizes existing results on reconciliation feasibility of binary gene trees and is thus an important step towards a full reconciliation algorithm for non-binary gene trees under the DLC model.

2   Background

2.1   Reconciliation Feasibility

We previously studied reconciliation feasibility under the DLC model [26] and review that work here.

We start with some basic tree and graph definitions. Let T be an unrooted, full, binary tree[1] with a set $V(T)$ of nodes (or vertices) and a set $E(T)$ of branches (or edges). Let $L(T) \subset V(T)$ denote the set of leaves, and for nodes u and v, let $path(u, v)$ denote the set of branches along the unique simple path from u to v in T. Similarly, let $\mathcal{G} = (V(\mathcal{G}), E(\mathcal{G}))$ be an undirected graph with a set $V(\mathcal{G})$ of vertices and a set $E(\mathcal{G})$ of edges. Let $\mathcal{C}(\mathcal{G})$ denote the set of connected components of $\mathcal{G}$, where $C \in \mathcal{C}(\mathcal{G})$ is a subgraph of $\mathcal{G}$ denoting a single connected component.

A *species tree* S is a tree that depicts the evolutionary history of a set of species, and a *gene tree* G is a tree that depicts the evolutionary history of a set of genes sampled from these species. Gene trees may be either binary or non-binary while the species tree is always assumed to be binary. A *species leaf map* $Le : L(G) \rightarrow L(S)$ associates each leaf of G with the leaf of S in which that gene is found. Note that more than one gene may be sampled from the same species; these genes could correspond to either multiple loci or multiple haploid samples. A gene tree is associated with a finite *locus set* $\mathbb{L}$ of species-specific loci that have evolved within the gene family. A *locus leaf map* $Le^L : L(G) \rightarrow \mathbb{L}$ associates each leaf of G with the species-specific locus at which that gene is found. For example, two genes map to the same species-specific locus if they are mapped to the same location on a reference genome. Note that the relationship between loci in different species is assumed to be unknown. Furthermore, there may exist copy number variations resulting in different samples from the same species containing different loci.

The *labeled coalescent tree (LCT)* formalizes the notion of a reconciliation in the DLC model [32]. In brief, the LCT is an annotated gene tree that simultaneously describes the gene tree topology and its reconciliation to the species tree. As a full description of the LCT is not necessary to characterize the reconciliation *feasibility* problem, we present only the necessary concepts and terminology. First, duplications occur along branches in the LCT, denoting that the locus has changed at some point along the branch. Second, the LCT labels each node and

---

[1] Branch lengths are not used in this work, so a tree always refers to a tree topology.

branch with the locus in which the gene evolves; for branches with a duplication, one side of the branch (before the duplication) is labeled with the original locus and the other side (after the duplication) with the new locus.

Multiple species-specific loci may be related through speciation events alone and thus correspond to the same evolutionary locus. This notion is formalized and used to define reconcilable gene trees as follows:

Definition 1 (Locus Class). *Let a collection* $\mathbb{LC} = \{C_i\}$ *of nonempty sets form a partition over* $\mathbb{L}$ *such that each locus* $l \in \mathbb{L}$ *belongs to a single* locus class $C_i \in \mathbb{LC}$.

Definition 2 (Reconcilable Gene Tree[2]). *Given gene tree* G*, species leaf map* Le*, and locus leaf map* Le^L*,* G *is said to be* reconcilable *if there exists some map* $\mathcal{L} \colon L(G) \cup E(G) \to \mathbb{LC}$ *of each leaf and edge of the gene tree to a single locus class, such that for each pair of genes* $g_1 \in L(G), g_2 \in L(G), g_1 \neq g_2$*,* $\mathcal{L}$ *is subject to the following constraints:*

1. *If* $Le^L(g_1) = Le^L(g_2)$*, then* $\mathcal{L}(g_1) = \mathcal{L}(g_2)$ *and for each* $e \in path(g_1, g_2)$*,* $\mathcal{L}(e) = \mathcal{L}(g_1)$*. (Allele Constraint)*
2. *If* $Le(g_1) = Le(g_2)$ *but* $Le^L(g_1) \neq Le^L(g_2)$*, then* $\mathcal{L}(g_1) \neq \mathcal{L}(g_2)$*. (Paralog Constraint)*

Constraint 1 ensures that genes from the same species-specific locus are assigned the same locus class and, because duplications create a unique new locus, that genes and edges assigned the same locus class form a subtree of the gene tree. Constraint 2 ensures that genes from paralagous loci are assigned different locus classes. Note that reconcilability of the gene tree depends on its topology and the mapping of its leaves to the leaves of the species tree and to species-specific loci, but reconcilability does not depend on the actual topology of the species tree.

Problem 3 (Reconciliation Feasibility). Given gene tree G, species leaf map Le, and locus leaf map Le^L, determine whether G is reconcilable.

The reconciliation feasibility problem can be solved using two structures, the Partially Labeled Coalescent Tree (PLCT, Fig. 1B) and the Locus Equivalence Graph (LEG, Fig. 1C), defined formally below:

Definition 4 (Partially Labeled Coalescent Tree). *Let* $\mathbb{P}(\mathbb{L})$ *denote the power set of* $\mathbb{L}$*. Given* G *and* Le^L*, the* partially labeled coalescent tree (PLCT) *is a map* $\mathcal{P} : E(G) \to \mathbb{P}(\mathbb{L})$ *constructed as follows: Consider each pair of genes* $g_1 \in L(G), g_2 \in L(G), g_1 \neq g_2$ *such that* $Le^L(g_1) = Le^L(g_2) = l$*. For each gene tree edge* $e \in path(g_1, g_2)$*, add* l *to* $\mathcal{P}(e)$*.*

Definition 5 (Locus Equivalence Graph). *Given a PLCT* $\mathcal{P}$ *for* G *and* Le^L*, the* locus equivalence graph (LEG) *is a graph* $\mathcal{G}$ *constructed as follows: Set* $V(\mathcal{G}) = \mathbb{L}$*. For each gene tree edge* $e \in E(G)$ *and each pair of loci* $l_1 \in \mathcal{P}(e), l_2 \in \mathcal{P}(e), l_1 \neq l_2$*, add* (l_1, l_2) *to* $E(\mathcal{G})$*.*

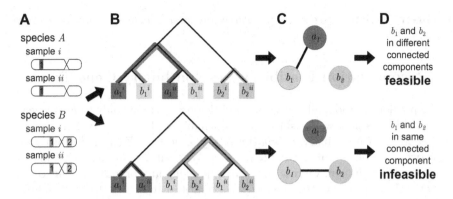

Fig. 1. Reconciliation feasibility for binary gene trees. (**A**) The sampled species (capital letters), loci (numbers), and haploid samples (roman numerals). We assume knowledge of the species-specific locus from which each gene is sampled. Within a species, genes at the same locus (across multiple samples must be alleles), and genes at different loci (regardless of sample) must be paralogs. (**B**) For a gene tree (black), the PLCT uses alleles to label branches along which no duplications are allowed (colored lines). (**C**) The LEG contains one node per species-specific locus and encodes overlapping labels in the PLCT as edges in the LEG. (**D**) A gene tree has a feasible reconciliation if and only if every connected component of the LEG contains no paralogs, that is, no more than one locus from each species. [Figure and caption adapted with permission from Rogers et al. [26].]

The PLCT captures the allele constraints for each species-specific locus by labeling edges of the gene tree with the species-specific locus or loci to which the edge must belong. If an edge is labeled with multiple loci, these multiple loci must correspond to the same locus class. This equivalency constraint is captured as an edge between loci in the LEG. Rogers et al. [26] provide a formal description of the algorithm for constructing the PLCT and LEG, describe an optimization, and derive their time complexities of $O(nk)$ and $O(nk^2)$, respectively, where $n = |L(G)|$ and $k = |\mathbb{L}|$. Next, paralog constraints are used to define reconcilable LEGs:

Definition 6 (Reconcilable Locus Equivalence Graph). *For each $l \in \mathbb{L}$, let map $Le^S : \mathbb{L} \to L(S)$ associate each species-specific locus with the leaf of S in which the locus is found. That is, for each $g \in L(G)$, if $l = Le^L(g)$, then $Le^S(l) = Le(g)$. Given G, Le, and Le^L, a LEG $\mathcal{G}$ for G and Le^L is said to be reconcilable if for each $C \in \mathcal{C}(\mathcal{G})$ and for each $s \in L(S)$, there exists no more than one locus $l \in C$ such that $Le^S(l) = s$.*

The LEG enforces the paralog constraints for each species by requiring that each connected component contain no more than one locus from any species. LEG reconcilability can be determined in $O(k^3)$ time and related to gene tree reconcilability [26]:

Theorem 7. *A gene tree is reconcilable if and only if its locus equivalence graph is reconcilable.*

3 Reconciliation Feasibility for Non-binary Gene Trees

In the previous section, all definitions and theorems applied only to *binary* gene trees. In this section, we consider the reconcilability of non-binary gene trees.

Let M be a non-binary, or *multifurcating*, gene tree. Each node with more than two children is called a *multifurcation*. Without loss of generality, and to simplify our discussion, we root M arbitrarily along any branch. A *binarization* $B(M)$ of M is a binary tree in which each multifurcation v with $k > 2$ children is replaced by a binary tree rooted at v with k leaves. These k leaves represent the k original children of v and thus may themselves be the roots of subtrees with their own descendants. The binary tree rooted at v is said to *resolve* the multifurcation, and we call that binary tree an *expansion tree* for v.

We now formalize the notion of reconcilable multifurcating gene trees:

Definition 8. *A multifurcating gene tree M is said to be* reconcilable *if there exists a binarization $B(M)$ of M that is reconcilable.*

Note that for a multifurcating gene tree, not all binarizations may be reconcilable. For example, two binarizations may induce different paths between two genes such that allele and paralog constraints are satisfiable in one binarization but not in another (Fig. 2). Rather than enumerate all binarizations and evaluate each for reconcilability, we propose to evaluate the reconcilability of multifurcating gene trees directly.

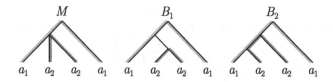

Fig. 2. Reconciliation feasibility for non-binary gene trees. A multifurcating gene tree M and two binarizations B_1 and B_2. Superscripts indicating haploid samples have been omitted. For B_1, a_1 and a_2 can be mapped to distinct locus classes, so the gene tree is reconcilable. For B_2, a_1 and a_2 must be mapped to the same locus class, but a_1 and a_2 are paralogs, so the gene tree is irreconcilable.

We start by applying the definitions of the PLCT and LEG (Definitions 4 and 5) directly to multifurcating gene trees. However, Theorem 7, which relates reconcilability of gene trees to reconcilability of LEGs requires that the gene tree be binary.[2] Our goal is to extend Theorem 7 to multifurcating gene trees:

---

[2] The proof considers only the single unique path between two genes in a binary tree.

Theorem 9. *A multifurcating gene tree is reconcilable if and only if its locus equivalence graph is reconcilable.*

For a multifurcating gene tree M, let the associated LEG be $\mathcal{G}_M$. We then reformulate Theorem 9 as two separate theorems, one for each direction of the "if and only if" statement:

Theorem 9a. *If $\mathcal{G}_M$ is reconcilable, then there exists a binarization $B(M)$ of M that is reconcilable.*

Theorem 9b. *If $\mathcal{G}_M$ is irreconcilable, then there exists no binarization $B(M)$ of M that is reconcilable.*

3.1 Proof of Reconcilability

For each locus $l \in \mathbb{L}$, there exists a *locus tree*[3] which is the subtree of M whose leaves contain that locus. Let $r(M)$ denote the root of M, and for $u \in V(M), u \neq r(M)$, let the *parent edge* of u be the edge from u to its parent. Consider a node u and its parent edge e. If edge e is not used by any locus trees, then u is said to be *uncontained*. However, if one or more locus trees contain edge e, then, by definition, the set of those loci are in a single connected component C of $\mathcal{G}_M$, and we say that u is *contained* by C.

Given a non-binary tree M (Fig. 3A), we want to efficiently determine whether or not there exists a binarization $B(M)$ of M that is reconcilable. We propose the following binarization algorithm:

1. For each multifurcation $v \in V(M)$, partition its children by the connected components in $\mathcal{G}_M$ that contain them, placing uncontained children arbitrarily (Fig. 3B).
2. For each set in the partition, construct a *sub-expansion tree* by attaching all the children to the leaves of an arbitrary binary tree with the same number of leaves as children in the set (Fig. 3C).
3. Join all sub-expansion trees together with another arbitrary binary tree of appropriate size, called the *connecting tree*, by attaching the roots of the sub-expansion trees to the leaves of the connecting tree. This results in our expansion tree for v (Fig. 3C).

Constructing an expansion tree for each multifurcating node in M, in this way, results in our binarization $B(M)$. Note that since some aspects of the construction permit arbitrary decisions (e.g., placement of uncontained nodes, construction of the connecting tree), the resulting binarization is not unique. We now relate $\mathcal{G}_M$ for M with $\mathcal{G}_{B(M)}$ for $B(M)$.

Lemma 10. *Let v be a multifurcating vertex in M, and let $B(M)$ be a binarization constructed by our algorithm. In $B(M)$, if a locus tree L for locus l contains an edge in the connecting tree of v, then it contains the parent edge of v.*

---

[3] Note that this locus tree is distinct from the locus tree of Rasmussen and Kellis [25].

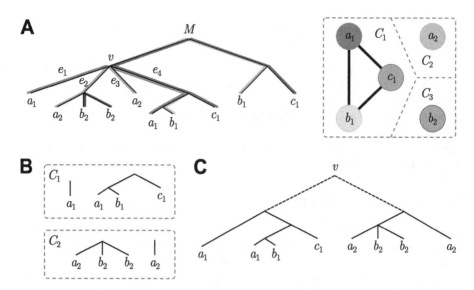

Fig. 3. Reconciliation feasibility for non-binary gene trees. (**A**) A multifurcating gene tree M and its locus equivalency graph $\mathcal{G}_M$. Superscripts indicating haploid samples have been omitted. Nodes with parent edges e_1 and e_4 are contained by connected component C_1, and nodes with parent edges e_2 and e_3 are contained by connected component C_2. (**B**) The partition of children from multifurcating node v. The first set includes the children contained by connected component C_1, and the second set includes the children contained by connected component C_2. (**C**) Sub-expansion trees (solid) joined through a connecting tree (dashed) to yield an expansion tree for v.

Proof. Suppose L contains an edge in the connecting tree of v but does not contain the parent edge of v. Then, by construction of $B(M)$, L has leaves g_1 and g_2 in two distinct sub-expansion trees of v. Therefore, in the original tree M, the path from g_1 to g_2 passes through v and thus passes through two children of v, denoted v_i and v_j. Since g_1 and g_2 are in distinct sub-expansion trees, it follows that v_i and v_j are each contained by a distinct connected component in $\mathcal{G}_M$. But, by definition of $\mathcal{G}_M$, the path from g_1 to g_2 implies that v_i and v_j are covered by a single connected component in $\mathcal{G}_M$. □

Lemma 11. *Let T be any binarization of M. Let l be a locus and let L_M and L_T be the locus trees for l in M and T, respectively. The edge set of L_T is exactly the edge set of L_M, with the addition of a subset of edges from expansion trees.*

Proof. Note that L_M is the union of paths between all pairs of leaves with locus l in M, and similarly, L_T is the union of paths between all pairs of leaves with locus l in T. Every path in M corresponds to a unique path in T where all internal nodes are expanded into a path through the corresponding expansion tree. By the uniqueness of paths in trees, L_T is exactly L_M augmented with the edges traversed in the expansion trees. □

For graph $\mathcal{G}$ and two nodes $u, v \in V(\mathcal{G})$, we say that u and v are *connected* if they are in the same connected component and *disconnected* otherwise.

Lemma 12. *Let l and k be a pair of disconnected loci in $\mathcal{G}_M$. Then, there is no edge between l and k in $\mathcal{G}_{B(M)}$.*

Proof. Let $T = B(M)$. Consider any l and k in different connected components in $\mathcal{G}_M$ and any multifurcation v in M. Let L_M and K_M denote the locus trees for l and k, respectively, in M, and let L_T and K_T denote the corresponding locus trees in T.

Since l and k are in different connected components of $\mathcal{G}_M$, at least one of L_M or K_M does not use the parent edge of v. By Lemma 11, the edge set of L_M and K_M are subsets of L_T and K_T, respectively. Therefore, at least one of L_T or K_T does not use the parent edge of v. Therefore, by Lemma 10, at most one of L_T or K_T contains an edge in the connecting tree for v in T.

Next, we claim that if l and k are in different connected components C_l and C_k in $\mathcal{G}_M$, then L_T and K_T do not intersect in any sub-expansion tree in T. Suppose L_T and K_T intersect inside a sub-expansion tree of some vertex v. Since C_l and C_k are in different components in $\mathcal{G}_M$, this intersection must happen at an edge that was introduced when joining the children of v into sub-expansion trees; these edges correspond to edges from v to its children in M. Thus, in M, L_M and K_M must share an edge and are thus in the same component, which contradicts our assumption.

We have established that if l and k are in different connected components in $\mathcal{G}_M$, then they cannot share an edge in either a connection tree or a sub-expansion tree in T. Therefore, by Lemma 11, l and k cannot share any edge in T and thus there is no edge between them in $\mathcal{G}_T$. $\qquad\square$

Finally, we prove Theorem 9a, which has been restated using $B(M)$ constructed by our algorithm.

Theorem 9a. *If $\mathcal{G}_M$ is reconcilable, then $\mathcal{G}_{B(M)}$ is reconcilable.*

Proof. Let $T = B(M)$. It suffices to show that if l and k are in different connected components in $\mathcal{G}_M$, then they are in different connected components in $\mathcal{G}_T$, implying that if $\mathcal{G}_M$ is reconcilable, then $\mathcal{G}_T$ is reconcilable.

Assume by way of contradiction that loci l and k are disconnected in M but connected in T. Then, there exists a path p from ℓ to k in $\mathcal{G}_T$. Let (u, v) be the first edge on p such that l and u are in the same connected component in $\mathcal{G}_M$ but u and v are in different connected components in $\mathcal{G}_M$. From Lemma 12, (u, v) cannot be an edge in $\mathcal{G}_T$, contradicting the assumption. $\qquad\square$

Theorem 9a implies a polynomial-time algorithm for *both* determining if a non-binary gene tree is reconcilable *and*, if so, constructing one reconcilable binarization. Recall that, for $n = |L(G)|$ and $k = |\mathbb{L}|$, it takes $O(nk) + O(nk^2) + O(k^3)$ time to construct the PLCT and LEG and then test the LEG for reconcilability. For the binarization process, let c denote the maximum number of children over all multifurcations in the gene tree, and m denote the total number of multifurcations. The time required to build the expansion tree for each multifurcation is linear in c, and each multifurcation can be resolved independently. Thus, the total complexity of the binarization process is $O(cm)$.

For comparison, the number of distinct binary resolutions for multifurcation v with k_v children is $N_v = (2k_v - 3)!!$. A brute-force approach that enumerates each binarization and combines them would therefore result in $\prod_{v \in V(M):k_v>2} N_v$ expansion trees, making it infeasible to enumerate and test each one for feasibility.

3.2   Proof of Irreconcilability

Theorem 9b. *If $\mathcal{G}_M$ is irreconcilable, then there exists no binarization $B(M)$ of M that is reconcilable.*

Proof. Let T be an arbitrary binarization of M. By Lemma 11, any pair of locus trees L_T and K_T in T contain all the edges of the corresponding locus tree, L_M and K_M, in M. Thus, any two loci that are connected by an edge in $\mathcal{G}_M$ must also have an edge in $\mathcal{G}_T$. □

It is not difficult to show that there exist non-binary gene trees that are not reconcilable.[4]

4   Discussion

We have presented an efficient algorithm that evaluates an arbitrary gene tree topology with an arbitrary number of samples per species under the DLC model and either (1) determines that there is a valid reconcilable binary resolution of that tree and constructs one such resolution or (2) determines that there exists no valid reconcilable binary resolution of that tree.

In previous work [26], we reconstructed RAxML [28] gene trees, collapsed poorly-supported branches to yield non-binary gene trees, and analyzed the reconcilability of the associated LEG. Our work here allows us to directly relate LEG reconcilability to gene tree reconcilability. In particular, while gene tree reconcilability is affected by poorly-supported branches, even multifurcating gene trees with well-supported branches can be infeasible.

One limitation of our work is that given a non-binary gene tree, we are not guaranteed to construct an *optimal* binary resolution. That is, our binarization may not yield a gene tree with the lowest reconciliation cost under a parsimony framework. But our work suggests one possible approach. We propose to explore the space of reconcilable resolutions compared to the space of all resolutions. If, in-practice, most non-binary gene trees have a single or small number of

---

[4] For example, in Fig. 3, swapping leaves labeled a_2 with leaves labeled c_1 would result in an irreconcilable LEG and thus a multifurcating gene tree for which there exists no reconcilable binarization.

reconcilable resolutions, it would imply that we could simply enumerate the resolutions, then apply existing reconciliation algorithms for binary trees.[5]

For irreconcilable gene trees, a possible research direction is to investigate error-correction algorithms. Such an algorithm could find the minimum number of topological rearrangements needed to yield a reconcilable gene tree. An alternative is to remove the minimum number of sampled individuals and explore possible patterns among the removed individuals. Such patterns could provide insight into whether certain populations are correlated with error and therefore more susceptible to problems elsewhere in a phylogenomic pipeline.

Acknowledgments. This work was supported by funds from the Department of Computer Science and the Dean of Faculty of Harvey Mudd College and by the U.S. National Science Foundation under grant IIS-1419739.

References

1. Åkerborg, Ö., Sennblad, B., Arvestad, L., Lagergren, J.: Simultaneous Bayesian gene tree reconstruction and reconciliation analysis. Proc. Nat. Acad. Sci. USA **106**(14), 5714–5719 (2009)
2. Albalat, R., Cañestro, C.: Evolution by gene loss. Nat. Rev. Genet. **17**, 379 (2016)
3. Bansal, M.S., Alm, E.J., Kellis, M.: Efficient algorithms for the reconciliation problem with gene duplication, horizontal transfer and loss. Bioinformatics **28**(12), i283–i291 (2012)
4. Burleigh, J.G., Bansal, M.S., Eulenstein, O., Hartmann, S., Wehe, A., Vision, T.J.: Genome-scale phylogenetics: inferring the plant tree of life from 18,896 gene trees. Syst. Biol. **60**(2), 117–125 (2011)
5. Chang, W.-C., Eulenstein, O.: Reconciling gene trees with apparent polytomies. In: Chen, D.Z., Lee, D.T. (eds.) COCOON 2006. LNCS, vol. 4112, pp. 235–244. Springer, Heidelberg (2006). https://doi.org/10.1007/11809678_26
6. Chauve, C., Doyon, J.P., El-Mabrouk, N.: Gene family evolution by duplication, speciation, and loss. J. Comput. Biol. **15**(8), 1043–1062 (2008)
7. Chen, K., Durand, D., Farach-Colton, M.: NOTUNG: a program for dating gene duplications and optimizing gene family trees. J. Comput. Biol. **7**(3–4), 429–447 (2000)
8. Chen, Z.Z., Deng, F., Wang, L.: Simultaneous identification of duplications, losses, and lateral gene transfers. IEEE/ACM Trans. Comput. Biol. Bioinform. **9**(5), 1515–1528 (2012)
9. Conow, C., Fielder, D., Ovadia, Y., Libeskind-Hadas, R.: Jane: a new tool for the cophylogeny reconstruction problem. Algorithm. Mol. Biol. **5**(16) (2010). https://almob.biomedcentral.com/articles/10.1186/1748-7188-5-16
10. David, L.A., Alm, E.J.: Rapid evolutionary innovation during an archaean genetic expansion. Nature **469**(7328), 93–96 (2011)
11. Degnan, J.H., Rosenberg, N.A.: Gene tree discordance, phylogenetic inference and the multispecies coalescent. Trends Ecol. Evol. **24**(6), 332–340 (2009)

---

[5] While most reconciliation algorithms do not support multiple samples per species nor non-binary gene trees, the former extension is fairly straightforward while the latter requires new algorithms.

12. Delsuc, F., Brinkmann, H., Philippe, H.: Phylogenomics and the reconstruction of the tree of life. Nat. Rev. Genet. **6**(5), 361–375 (2005)
13. Doyon, J.-P., Scornavacca, C., Gorbunov, K.Y., Szöllősi, G.J., Ranwez, V., Berry, V.: An efficient algorithm for gene/species trees parsimonious reconciliation with losses, duplications and transfers. In: Tannier, E. (ed.) RECOMB-CG 2010. LNCS, vol. 6398, pp. 93–108. Springer, Heidelberg (2010). https://doi.org/10.1007/978-3-642-16181-0_9
14. Goodman, M., Czelusniak, J., Moore, G.W., Romero-Herrera, A., Matsuda, G.: Fitting the gene lineage into its species lineage, a parsimony strategy illustrated by cladograms constructed from globin sequences. Syst. Zool. **28**(2), 132–163 (1979)
15. Górecki, P., Tiuryn, J.: Dls-trees: a model of evolutionary scenarios. Theor. Comput. Sci. **359**(1–3), 378–399 (2006)
16. Koonin, E.V.: Orthologs, paralogs, and evolutionary genomics. Annu. Rev. Genet. **39**(1), 309–338 (2005)
17. Kordi, M., Bansal, M.S.: Exact algorithms for duplication-transfer-loss reconciliation with non-binary gene trees. IEEE/ACM Trans. Comput. Biol. Bioinform. **PP**(99), 1 (2018)
18. Lafond, M., Swenson, K.M., El-Mabrouk, N.: An optimal reconciliation algorithm for gene trees with polytomies. In: Raphael, B., Tang, J. (eds.) WABI 2012. LNCS, vol. 7534, pp. 106–122. Springer, Heidelberg (2012). https://doi.org/10.1007/978-3-642-33122-0_9
19. Maddison, W.P.: Gene trees in species trees. Syst. Biol. **46**(3), 523–536 (1997)
20. Ochman, H.: Lateral and oblique gene transfer. Curr. Opin. Genet. Dev. **11**(6), 616–619 (2001)
21. Ohno, S.: Evolution by Gene Duplication. Springer, Heidelberg (1970). https://doi.org/10.1007/978-3-642-86659-3
22. Page, R.D.: Maps between trees and cladistic analysis of historical associations among genes, organisms, and areas. Syst. Biol. **43**(1), 58–77 (1994)
23. Rannala, B., Yang, Z.: Bayes estimation of species divergence times and ancestral population sizes using DNA sequences from multiple loci. Genetics **164**(4), 1645–1656 (2003)
24. Rasmussen, M.D., Kellis, M.: A Bayesian approach for fast and accurate gene tree reconstruction. Mol. Biol. Evol. **28**(1), 273–290 (2011)
25. Rasmussen, M.D., Kellis, M.: Unified modeling of gene duplication, loss, and coalescence using a locus tree. Genome. Res. **22**, 755–765 (2012)
26. Rogers, J., Fishberg, A., Youngs, N., Wu, Y.C.: Reconciliation feasibility in the presence of gene duplication, loss, and coalescence with multiple individuals per species. BMC Bioinform. **18**, 292 (2017)
27. Slowinski, J.B.: Molecular polytomies. Mol. Phylogenet. Evol. **19**(1), 114–120 (2001)
28. Stamatakis, A.: RAxML-VI-HPC: maximum likelihood-based phylogenetic analyses with thousands of taxa and mixed models. Bioinformatics **22**(21), 2688–2690 (2006)
29. Tofigh, A., Hallett, M., Lagergren, J.: Simultaneous identification of duplications and lateral gene transfers. IEEE/ACM Trans. Comput. Biol. Bioinform. **8**(2), 517–535 (2011)
30. Vilella, A.J., Severin, J., Ureta-Vidal, A., Heng, L., Durbin, R., Birney, E.: Ensemblcompara genetrees: complete, duplication-aware phylogenetic trees in vertebrates. Genome. Res. **19**(2), 327–335 (2009)

31. Wu, T., Zhang, L.: Structural properties of the reconciliation space and their applications in enumerating nearly-optimal reconciliations between a gene tree and a species tree. BMC Bioinform. **12**(Suppl 9), S7 (2011)

32. Wu, Y.C., Rasmussen, M.D., Bansal, M.S., Kellis, M.: Most parsimonious reconciliation in the presence of gene duplication, loss, and deep coalescence using labeled coalescent trees. Genome Res. **24**(3), 475–486 (2014)

33. Zhang, L.: From gene trees to species trees ii: species tree inference by minimizing deep coalescence events. IEEE/ACM Trans. Comput. Biol. Bioinform. **8**(6), 1685–1691 (2011)

34. Zheng, Y., Zhang, L.: Reconciliation with non-binary Gene trees revisited. In: Sharan, R. (ed.) RECOMB 2014. LNCS, vol. 8394, pp. 418–432. Springer, Cham (2014). https://doi.org/10.1007/978-3-319-05269-4_33

35. Zmasek, C.M., Eddy, S.R.: A simple algorithm to infer gene duplication and speciation events on a gene tree. Bioinformatics **17**(9), 821–828 (2001)

Solving the Tree Containment Problem for Reticulation-Visible Networks in Linear Time

Andreas D. M. Gunawan[(⊠)]

Department of Mathematics, National University of Singapore,
10 Lower Kent Ridge Road, Singapore, Singapore
a0054645@nus.edu.sg

Abstract. The tree containment problem (TCP) is a fundamental problem in phylogenetic study. It was introduced as a mean for verifying whether a network is consistent with a binary tree. The containment problem is NP-complete, even if the network input is binary. If the input is restricted to reticulation-visible networks, the TCP has been proved to be solvable in quadratic time. In this paper, we show that there is a linear time TCP algorithm for binary reticulation-visible networks.

Keywords: Phylogeny reconstruction · Tree containment problem
Reticulation-visible

1 Introduction

Binary trees have been used to model evolutionary history for a long time. In a binary tree, an internal node represents a speciation event (i.e. the emerging of a new species) whereas a leaf represents an existing species. Recently, researchers discover reticulate evolutionary events such as hybridization and horizontal gene transfer [3,10], where genetic material may flow from one species to another. As these reticulation events cannot be explained using trees, researchers develop a more general model called phylogenetic networks (or simply networks), where the reticulation events are represented by internal nodes of indegree more than one.

Although binary tree model cannot explain reticulation events, it is still widely used due to its simplicity. For instance, by comparing similarity between homologous proteins across different species, biologists can produce a binary tree that can best predict the evolution. One way to verify whether a binary tree is consistent with a phylogenetic network is by checking whether the latter contains a subtree that is conceptually the same as the former. This verification problem is known as the tree containment problem (TCP), and is known to be NP-complete, even if the network input is restricted to binary networks [8].

J. Jansson et al. (Eds.): AlCoB 2018, LNBI 10849, pp. 24–36, 2018.
https://doi.org/10.1007/978-3-319-91938-6_3

In order to make the network model practical, much effort has been devoted to find big network classes where the TCP can be solved quickly. In [7], van Iersel et al. posed the question whether the TCP is solvable in polynomial time for the class of binary reticulation-visible networks. Positive answers were independently obtained five years after in [2,4], the former gave a cubic time algorithm while the latter gave a quadratic time algorithm. Since then, the class of reticulation-visible networks remains to be one of the largest network classes where the TCP is solvable in polynomial time.

In this paper, we present a TCP algorithm for binary reticulation-visible networks that runs in linear time. We use a decomposition theorem that was introduced in [4], which has also been used to produce two fast algorithms (albeit exponential) for solving the TCP [5] and a related network verification problem called the cluster containment problem [9] for arbitrary networks.

2   Basic Definitions and Notations

A (phylogenetic) *network* is a rooted directed acyclic graph that satisfies: (i) a non-root node is of either indegree or outdegree one, and (ii) the leaves (i.e. nodes of outdegree zero) are bijectively labeled by a set of taxa of some species under study. A non-leaf node is called a *tree node* if it is of indegree one, and otherwise is called a reticulation node, or simply *reticulation*. Of note, a reticulation is of outdegree one by property (i) above. For convenience, we add an incoming edge with open end to the root, thereby making it a tree node.

For a given network N, ρ_N denotes its root, $\mathcal{V}(N)$ its set of nodes, $\mathcal{E}(N)$ its set of edges, $\mathcal{T}(N)$ its set of tree nodes (including the root), $\mathcal{R}(N)$ its set of reticulations, and $\mathcal{L}(N)$ its set of leaves. A *binary network* is a network where every internal node is of total degree three. Of note, a binary tree is a binary phylogenetic network that has no reticulation.

An incoming edge of a reticulation (resp. a tree node or a leaf) is called a *reticulation edge* (resp. *tree edge*). A *tree path* is a path consisting of only tree edges.

Node u is a *parent* of node v (or v is a *child* of u) if the edge (u,v) exists. Two nodes are *siblings* if they have a common parent. In a more general context, node u is *above* node v (or u is an *ancestor* of v) if there is a path from u to v. The subnetwork of N induced by the nodes below v is denoted as $N[v]$.

A node u is a *visible ancestor* of another node v if every path from the network root to v passes u. A node is *visible* if it is the visible ancestor of some labeled leaf in N; otherwise it is invisible. A *reticulation-visible* network is a network where every reticulation is visible. It is worth noting that if N is a binary reticulation-visible network with n leaves, then there are at most $8n - 7$ nodes and $11n - 10$ edges [2,6].

For a set of nodes V and a set of edges E, $N - V - E$ is the network with node set $\mathcal{V}(N) \backslash V$ and edge set $\{(u,v) \in \mathcal{E}(N) \backslash E : u, v \notin V\}$. If $V \cup E$ comprises a single element x, we may simply write the resulting network as $N - x$.

2.1 Tree Containment Problem

Let u be a node in a network N. If u has only one parent v and one child w, then suppressing u means we replace u and all edges adjacent to it with an edge (i.e. we update the network into $N - u + (v, w)$). If u is a dummy leaf (i.e. a leaf that is not labeled with any taxa), then suppressing u means we remove the node u from N. Of note, a dummy leaf may exist after removing some edges from a network. From now on, nodes of total degree two and dummy leaves are called *redundant nodes*.

Let G be a binary tree, and let N be a network. A subtree T of N is a *subdivision* of G if we can get G from T by repeatedly suppressing redundant nodes. The tree containment problem (TCP; see Fig. 1 for illustration) is then formulated as follows:

Tree containment problem (TCP)
Instance: A phylogenetic network N and a full binary tree G over the same set of leaves.
Question: Does N display G?

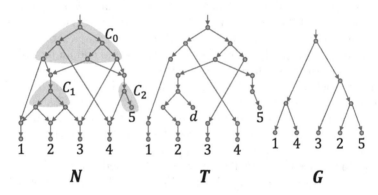

Fig. 1. A binary reticulation-visible network N (**left**) which contains three nontrivial components (namely, C_0, C_1, C_2) and four trivial components ($\{1\}, \{2\}, \{3\},$ *and* $\{4\}$). The components C_1 and C_2 are exposed. N contains a spanning subtree T (**middle**) that is a subdivision of the binary tree G (**right**), therefore N displays G. The node d in T is a dummy leaf

2.2 A Decomposition Theorem

Removing every reticulation from N yields a forest $N - \mathcal{R}(N)$, which comprises the tree nodes and leaves in N. Each maximal connected component in this forest is called a tree node component (or simply *component*) of N. The root of a component C is the topmost tree node in the component, and is denoted as $\rho(C)$. See Fig. 1 for an illustration.

A tree component C is below another component C' if $\rho(C)$ is below $\rho(C')$. This induces a partial ordering on the components, and thus we can label the components with $C_0, C_1, \ldots, C_k$ such that $\rho(C_0) = \rho_N$ and C_j is below C_i only if $i < j$. A tree component is *trivial* if it contains only a single leaf. A nontrivial tree component is *exposed* if every tree component below it is trivial. The following decomposition theorem was established in [4]:

Theorem 1 (Network Decomposition). *Let $C_0, C_1, \ldots, C_q$ be the components of a reticulation-visible network N. The following statements are true:*

(i) *$\mathcal{T}(N) \cup \mathcal{L}(N) = \uplus_{i=0}^{q} \mathcal{V}(C_i)$, where $\uplus$ denotes union of disjoint sets.*

(ii) *The child of a reticulation is a component root, and each parent of a reticulation is a tree node that belongs to some component. Additionally, every component root is visible.*

(iii) *$|\mathcal{V}(C_i)| = 1$ if and only if C_i is trivial. If $|\mathcal{V}(C_i)| > 1$, then C_i contains either a network leaf or there is a reticulation whose parents are all in C_i.*

(iv) *If $|\mathcal{V}(N)| > 1$, there is at least one exposed component.*

3   The TCP Algorithm

Throughout this section, let N be a binary reticulation-visible network and G be a binary tree for the TCP. Similar as the algorithms in [4,5,9], the TCP algorithm in this paper also uses the decomposition theorem to approach the containment problem in a divide-and-conquer manner. First, we pre-process the input network N to decompose it into its components. We then pick a nontrivial exposed component and observe its structure. From this observation, we either deduce that the TCP has a negative answer or simplify the network by dissolving the exposed component (along with some corresponding subtree of G) into a single leaf. Finally, we repeat the process and find another exposed component. An illustration of the process can be found in Fig. 2.

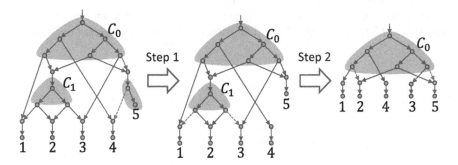

Fig. 2. An example of the divide-and-conquer approach in solving the TCP. In each step, we choose an exposed component, remove several reticulation edges (dotted red edges) from the reticulation below it, and contract the component into a single leaf (Color figure online)

3.1   Early Results

Let u be a node below ρ_C in N, and let x be a node in G. To simplify notation, we let $L_M(u)$ denote the set of (labeled) leaves below a node u in some network M. We also use $L_M^+(u)$ to denote the set of leaves below u of which u is a visible ancestor.

We introduce a Boolean variable $f(u,x)$ to indicate whether or not the subnetwork of N below u displays the subtree of G below x, that is:

$$f(u,x) = \begin{cases} 1 \text{ if } N[u] \text{ displays } G[x], \\ 0 \text{ otherwise.} \end{cases} \tag{1}$$

We remark here that for $N[u]$ to display $G[x]$, we do not impose an additional requirement that $\ell \notin L_N^+(u)$ for every leaf ℓ not below x in G like what we did in [4]. For every pair or leaves u in N and x in G, clearly we have $f(u,x) = 1$ if and only if $x = u$ (i.e. they are labeled by the same taxa). The following proposition can then be used to compute $f(u,x)$:

Proposition 2 (Proposition 8, [4]). *For any tree node u in some exposed component of N, $f(u,x) = 1$ if and only if one of the following two conditions is true:*

(1) *u has a child v, such that $f(v,x) = 1$; or*
(2) *x has two children y, z and u has two distinct children v and w, such that $f(v,y) = 1$ and $f(w,z) = 1$.*

Now, suppose that u is a visible node (thus $L_N^+(u) \neq \emptyset$). For a fixed leaf $\ell_u \in L_N^+(u)$, we define $t(u,\ell_u)$ as:

$$t(u,\ell_u) = \max(\{x \in \mathcal{V}(G) : x \text{ is above } \ell_u \text{ and } f(u,x) = 1\}). \tag{2}$$

In words, $t(u,\ell_u)$ is the highest ancestor of ℓ_u in G, such that the subtree of G below $t(u,\ell_u)$ is displayed by the network $N[u]$.

We remark that the node v_{d_C} defined in [4] is the same as $t(\rho_C, \ell_{\rho_C})$ for some $\ell_{\rho_C} \in L_N^+(\rho_C)$, so the above definition can be seen as a generalization of the result in the previous paper. Additionally, Propositions 4 and 5 in the same paper can be generalized as follows (the proofs are similar and thus omitted):

Proposition 3. *If N displays G, then the node $t(u, \ell_u)$ is above every leaf in G that is also in $L_N^+(u)$.*

Proposition 4. *If N displays G, then G has a subdivision T that is a subtree of N such that $t(u, \ell_u)$ in G corresponds to some node below u in T.*

Proposition 3 can be used to deduce that the TCP has negative answer if $L_N^+(u) \not\subseteq L_G(t(u, \ell_u))$. We also remark that for some node u in an exposed component, every leaf $\ell \in L_N(u) \backslash L_N^+(u)$ has a reticulation parent r, such that r has two parents p and q that must be below and not below u, respectively. Therefore, Proposition 4 implies that we can remove the edge (q, r) if ℓ is below t in G (thereby keeping ℓ below u) without changing the outcome of the TCP. Similarly, we can remove (p, r) if ℓ is not below t in G. This can be followed by replacing $N[u]$ and $G[t(u, \ell_u)]$ with a leaf.

In order to dissolve an exposed component, in [4] we compute $v_{d_C} = t(\rho_C, \ell_{\rho_C})$ by first computing the value of $f(u, x)$ for every node u in C and x in G. This is in fact the biggest hurdle for breaking the quadratic time bound. In the next subsection (Lemma 7), we show how to compute $t(s, \ell_s)$ in linear time if s satisfies some special conditions. This allows us to dissolve the exposed component part by part, rather than dissolving the whole exposed component at once as in [4].

3.2 Computing the Largest Subtree Displayed Below Some Node with Certain Properties

Let C be an exposed component, and suppose that its root ρ_C is a visible ancestor of some leaf ℓ. As C is exposed, we have that either ℓ is a network leaf in C or the parent of ℓ is a reticulation whose parents are both in C. In the latter case, there is a unique node in C that is the lowest visible ancestor of the two grandparents of ℓ. Such a node is called the *split node* of ℓ, and is denoted as spl(ℓ).

Let spl(C) denote the set of split nodes in C, that is:

$$\mathrm{spl}(C) = \{\mathrm{spl}(\ell) : \ell \in L_N^+(\rho_C) \setminus \mathcal{V}(C)\}.$$

In addition, we define $\mathrm{spl}^*(C) = \mathrm{spl}(C) \cup \{\rho_C\}$. A node s is a *lowest* split node if $s \in \mathrm{spl}^*(C)$ and s has no strict descendant in $\mathrm{spl}^*(C)$. Such node satisfies the following lemma:

Lemma 5. *If s is one of the lowest nodes in $\mathrm{spl}^*(C)$, then s and every node below it satisfies the following key property:*
($\star$) for every child s' of s and reticulation r, r has at most one parent below s'.

Proof. If s satisfies condition $(\star)$, then for every node v strictly below s, $N[v]$ is only a tree, and therefore v also satisfies $(\star)$. Thus it is enough to prove that s satisfies $(\star)$.

If $\mathrm{spl}(C) = \emptyset$, then C has no split node and $s = \rho_C$. This implies that every reticulation below C has exactly one parent in C, and thus the proposition follows.

If $\mathrm{spl}(C)$ is not empty, then $s = \mathrm{spl}(\ell)$ for some leaf ℓ in a trivial component below C such that the two grandparents of ℓ are both in C. Let s' be any child of s. Suppose for contradiction that there is a reticulation r below s' whose parents are both below s' in C. As C is an exposed component, the child of r is a leaf ℓ'. But this implies that $\mathrm{spl}(\ell')$ is another split node below s', contradicting the fact that s is a lowest node in $\mathrm{spl}^*(C)$. This concludes the proof.   □

Of note, condition $(\star)$ implies that for every child s' of s, the subnetwork $N[s']$ is just a tree. Now, let $\mathrm{par}_M(v)$ and $\mathrm{sib}_M(v)$ be the parent and the sibling of v in network M, if any. The following lemma holds:

Lemma 6. *Let u be a node in C such that there is a tree path from u to a leaf ℓ_u. If $par_N(u)$ is a tree node, then*

$$t(par_N(u), \ell_u) = \begin{cases} par_G(x) \ if \ N[sib_N(u)] \ displays \ G[sib_G(x)], \ and \\ x \ otherwise, \end{cases}$$

where $x = t(u, \ell_u)$.

Proof. By definition of $t(\cdot, \cdot)$ in Eq. 2, we have: (i) $f(u, x) = 1$, and (ii) $f(u, z) = 0$ for any node z that is strictly above x. Moreover, there is a tree path from u to ℓ_u, which implies that the sibling of u is not above ℓ_u. This fact and the fact that x is an ancestor of ℓ_u implies: (iii) $f(\mathrm{sib}_N(u), z) = 0$ for any ancestor z of x. Then there are three observations that we can deduce from Proposition 2.

First, $f(\mathrm{par}_N(u), x) = 1$ as $f(u, x) = 1$ (condition (1) in Proposition 2).

Second, we claim $f(\mathrm{par}_N(u), \mathrm{par}_G(\mathrm{par}_G(x))) = 0$. Suppose otherwise for contradiction, then either condition (1) or (2) in Proposition 2 must hold. Condition (1) cannot hold, because (ii) implies that $f(u, \mathrm{par}_G(\mathrm{par}_G(x))) = 0$, and (iii) implies that $f(\mathrm{sib}_N(u), \mathrm{par}_G(\mathrm{par}_G(x))) = 0$. Consequently, condition (2) must hold, and thus either $[f(u, \mathrm{par}_G(x)) = 1$ and $f(\mathrm{sib}_N(u), \mathrm{sib}_G(\mathrm{par}_G(x))) = 1]$ or $[f(\mathrm{sib}_N(u), \mathrm{par}_G(x)) = 1$ and $f(u, \mathrm{sib}_G(\mathrm{par}_G(x))) = 1]$. The former contradicts (ii), whereas the latter contradicts (iii), and thus the claim holds.

Third, $f(\mathrm{par}_N(u), \mathrm{par}_G(x)) = 1$ if and only if $f(\mathrm{sib}_N(u), \mathrm{sib}_G(x)) = 1$. By Proposition 2, $f(\mathrm{par}_N(u), \mathrm{par}_G(x)) = 1$ if and only if one of the following holds:

(a) $f(u, \mathrm{par}_G(x)) = 1$ or $f(\mathrm{sib}_N(u), \mathrm{par}_G(x)) = 1$;
(b) $f(u, \mathrm{sib}_G(x)) = 1$ and $f(\mathrm{sib}_N(u), x) = 1$; or
(c) $f(u, x) = 1$ and $f(\mathrm{sib}_N(u), \mathrm{sib}_G(x)) = 1$.

Condition (a) contradicts (ii) or (iii), whereas (b) contradicts (iii), and thus the observation follows from (c).

The rest of the proof then follows from the three observations and the definition of $t(\cdot, \cdot)$ in Eq. 2.   □

Algorithm 1. A subroutine for finding $t(s, \ell_s)$ that follows from Lemma 7

Procedure t-FINDER(N, G, s, ℓ_s)
    Input: A binary reticulation-visible N, a full binary tree G, a visible node s
        that satisfies $(\star)$, and a leaf $\ell_s \in L_N^+(s)$.
    Output: A node $t(s, \ell)$ as defined in Eq. (2).

1   if ℓ_s *has a tree node parent* **then**
2   set $u = \ell_s$ in N and $t = \ell_s$ in G;
3   while $u \neq s$ **do**
4   if u *has a sibling* **then**
5   set $u' = \mathrm{sb}_N(u)$ and $t' = \mathrm{sb}_G(t)$;
6   if $N[u']$ *displays* $G[t']$ **then**
7                 $t = \mathrm{par}_G(t)$;

8   set $u = \mathrm{par}_N(u)$;

9   else
10   Let e_1, e_2 be the incoming edges of $\mathrm{par}_N(\ell_s)$;
11      $t_1 = t$-FINDER$(N - e_1, G, s, \ell_s)$;
12      $t_2 = t$-FINDER$(N - e_2, G, s, \ell_s)$;
13      $t = \max(t_1, t_2)$;

14   **return** t;

Empowered by Lemmas 5 and 6 above, we present the following lemma:

Lemma 7. *If s is an internal node in C such that s satisfies condition $(\star)$ and $\ell_s \in L_N^+(s)$, then Algorithm 1 correctly finds $t(s, \ell_s)$ in $O(|N[s]|)$ time.*

Proof. (**Correctness**) There are two possible cases. First, assume that ℓ_s has a tree node parent (thus ℓ_s is in C), in which case Lines 1–8 are executed (see Fig. 3 for illustration). Then there is a tree path:

$$P : u_{k+1} = s, u_k, \ldots, u_1 = \ell_s.$$

We then apply Lemma 6 recursively on u_i in ascending order for each $2 \leq i \leq k+1$. To find $t(u_i, \ell_s)$, we simply check whether the subnetwork of N below the sibling of u_{i-1} displays the subtree of G below the sibling of $t(u_{i-1}, \ell_s)$ (Line 6). Therefore we correctly obtain $t(s, \ell_s)$ when the procedure ends.

For the second case, assume that ℓ_s has a reticulation parent, such that the two grandparents of ℓ_s are in C. In this paragraph, we use $t_M(\cdot, \cdot)$ instead of $t(\cdot, \cdot)$ when considering a network M to avoid ambiguity. Let e_1, e_2 be the incoming edges of the reticulation parent of ℓ_s, and let $N_i = N - e_i$ for $i = 1, 2$. Clearly a subtree of $N[s]$ is a subtree of either $N_1[s]$ or $N_2[s]$ and vice versa; therefore we have

$$t_N(s, \ell_s) = \max(t_{N_1}(s, \ell_s), t_{N_2}(s, \ell_s)).$$

Computing $t_{N_1}(s, \ell_s)$ and $t_{N_2}(s, \ell_s)$ can be done by repeating the procedure in the first case (Lines 11 and 12), and this completes the proof of the correctness.

N[s]

G[v₃]

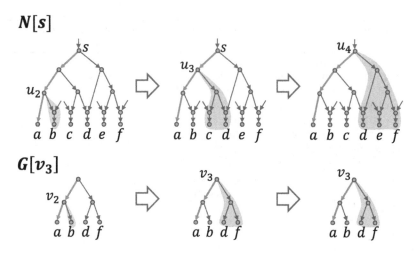

Fig. 3. An illustration of the proof of Lemma 7. The top figure shows a subnetwork of N rooted at s, whereas the bottom figure shows a subtree of G. The node s satisfies condition $(\star)$, and there is a tree path from s to the leaf a. Thus, following the notation in Lemma 7, we set $u_1 = a$ and $u_{i+1} = \text{par}_N(u_i)$ for $i = 1, 2, 3$. The nodes in G is labeled similarly, but with v_i instead of u_i. Initially, we set $t(u_1, a) = a$, and compare the subtree of N branching off u_2 with the subtree of T branching off v_2 (highlighted in grey background) to deduce that $t(u_2, a) = v_2$. Next, we compare the subtree of N branching off u_3 with the subtree of T branching off v_3. The subtree of G is not displayed here, so $t(u_3, a) = t(u_2, a) = v_2$. Finally, the subtree of G branching off v_3 is displayed in the subtree of N branching off u_4, so $t(s, a) = v_3$ and we complete the procedure

(Complexity) Consider a call of Line 6, where we need to check whether $N[u']$ displays $G[t']$. We first check whether $L_G(t') \subseteq L_N(u')$. This can be done in $O(|N[u']|)$ time by a breadth-first-search on N and G to find every leaf below u' and t' (we quit once we find a leaf below t' that is not below u'). If $L_G(t') \nsubseteq L_N(u')$, then we immediately obtain a negative answer, so we assume otherwise. We can then find a subtree of $N[u']$ over the same leaves as $L_G(t')$ and suppress every redundant node in it to obtain a binary subtree T, which can then be compared with $G[t']$. The construction of T can be done via a post-order traversal on $N[u']$, whereas the comparison of T and G can be done by a simultaneous post-order traversal on T and $G[t']$ to check if there is any difference between them. Thus, an execution of Line 6 can be done in $O(|N[u']|)$ time.

Now, suppose that ℓ_s has a tree node parent, and thus Lines 3–8 are executed. It is not hard to see that Lines 4, 5, 7, and 8 can be done in constant time. Therefore, an execution of the 'While' loop in Line 3 when considering the node u_i takes $O(N[\text{sib}_N(u_i)])$ by the previous paragraph. The 'While' loop at line 3 is executed for each $u_i, 1 \le i \le k$. Moreover, s satisfies condition $(\star)$, so $N[u_k]$ is a

component C. Lines 4 and 5 can be done with one post-order traversal on C, thus they require $O(|\mathcal{V}(N[\rho_C])|)$ time.

Clearly, the condition for the 'If'/'ElseIf' command in Lines 8, 9, and 12 can be checked in $O(1)$ time. Moreover, the computation of $l_{\min}$ and $r_{\max}$ also requires $O(1)$ time. As we need to repeat this for each node in C, Lines 7–16 require $O(|\mathcal{V}(C)|)$ time for each 'While' loop execution.

Lines 19 and 21 require $O(|\mathcal{V}(N[s])|)$ time. Line 20 also requires the same time by Lemma 7. The conditions on Lines 24–26 can each be checked in $O(1)$ time, so the 'Foreach' loop requires $O(|\mathcal{V}(C)|)$ time. In each execution of Line 27, we replace $N[s]$ by a single leaf. Therefore Lines 17–27 requires $O(|\mathcal{V}(N[\rho_C])|)$ time for each 'While' loop execution.

Hence an iteration of the 'While' loop requires $O(|\mathcal{V}(N[\rho_C])|)$ time for dissolving an exposed component C into a single leaf. As in each iteration we consider different exposed component, we therefore conclude this discussion with the following theorem:

Theorem 8. *Algorithm 2 correctly solves the TCP for a binary reticulation-visible network N in $O(|\mathcal{V}(N)|)$ time.*

4 Conclusion

We further improve the time complexity of TCP algorithm for binary reticulation-visible networks into linear time. However, this method relies heavily on the facts that the indegree of each reticulation r is two and there are no two adjacent reticulations. Without this assumption, there could be multiple "split node" for each reticulation and some may not be visible, which is a crucial property that is used here. Thus, it is unclear whether similar method can be applied to obtain a linear time TCP algorithm for arbitrary reticulation-visible network.

A similar result was independently obtained in [11], only a few days after we obtained this result. Although the algorithm in [11] is more general as it can be applied for non-binary case (in which case the running time might be super-linear), Weller's algorithm requires a pre-processing on the components which allows us to find the lowest common ancestors of any two nodes in $O(1)$ time (e.g. see [1]). Although linear in theory, such pre-processing is not easy to implement and may take some time to run. In contrast, our algorithm only requires some simple traversals, and thus is potentially faster.

Acknowledgement. I'd like to thank Prof. Zhang Louxin for his guidance and help. The project is financially supported by Singapore MoE-ARF Tier 1 grant, R-146-000-238-114.

References

1. Bender, M.A., Farach-Colton, M., Pemmasani, G., Skiena, S., Sumazin, P.: Lowest common ancestors in trees and directed acyclic graphs. J. Algorithms **57**(2), 75–94 (2005)
2. Bordewich, M., Semple, C.: Reticulation-visible networks. Adv. Appl. Math. **78**, 114–141 (2016)
3. Chan, J.M., Carlsson, G., Rabadan, R.: Topology of viral evolution. PNAS **110**(46), 18566–18571 (2013)
4. Gunawan, A.D.M., DasGupta, B., Zhang, L.: A decomposition theorem and two algorithms for reticulation-visible networks. Inf. Comput. **252**, 161–175 (2017)
5. Gunawan, A.D.M., Lu, B., Zhang, L.: A program for verification of phylogenetic network models. Bioinformatics **32**(17), i503–i510 (2016)
6. Gunawan, A.D.M., Zhang, L.: Bounding the size of a network defined by visibility property (2015). http://arxiv.org/abs/1510.00115
7. van Iersel, L., Semple, C., Steel, M.: Locating a tree in a phylogenetic network. Inf. Process. Lett. **110**(23), 1037–1043 (2010)
8. Kanj, I.A., Nakhleh, L., Than, C., Xia, G.: Seeing the trees and their branches in the network is hard. Theor. Comput. Sci. **401**, 153–164 (2008)
9. Lu, B., Zhang, L., Leong, H.W.: A program to compute the soft Robinson-Foulds distance between phylogenetic networks. BMC Genomics **18**(2), 111 (2017)
10. Marcussen, T., Sandve, S.R., Heier, L., Spannagl, M., Pfeifer, M., Jakobsen, K.S., Wulff, B.B.H., Steuernagel, B., Mayer, K.F.X., Olsen, O., et al.: Ancient hybridizations among the ancestral genomes of bread wheat. Science **345**(6194), 1250092 (2014)
11. Weller, M.: Linear-time tree containment in phylogenetic networks. arXiv preprint arXiv:1702.06364 (2017)

Polynomial-Time Algorithms
for Phylogenetic Inference Problems

Leo van Iersel[1]([✉]), Remie Janssen[1], Mark Jones[1], Yukihiro Murakami[1],
and Norbert Zeh[2]

[1] Delft Institute of Applied Mathematics, Delft University of Technology,
Van Mourik Broekmanweg 6, 2628 XE Delft, The Netherlands
{L.J.J.vanIersel,R.Janssen-2,M.E.L.Jones,Y.Murakami}@tudelft.nl
[2] Faculty of Computer Science, Dalhousie University,
6050 University Ave, Halifax, NS B3H 1W5, Canada
nzeh@cs.dal.ca

Abstract. A common problem in phylogenetics is to try to infer a species phylogeny from gene trees. We consider different variants of this problem. The first variant, called UNRESTRICTED MINIMAL EPISODES INFERENCE, aims at inferring a species tree based on a model of speciation and duplication where duplications are clustered in duplication episodes. The goal is to minimize the number of such episodes. The second variant, PARENTAL HYBRIDIZATION, aims at inferring a species *network* based on a model of speciation and reticulation. The goal is to minimize the number of reticulation events. It is a variant of the well-studied HYBRIDIZATION NUMBER problem with a more generous view on which gene trees are consistent with a given species network. We show that these seemingly different problems are in fact closely related and can, surprisingly, both be solved in polynomial time, using a structure we call "beaded trees". However, we also show that methods based on these problems have to be used with care because the optimal species phylogenies always have some restricted form. We discuss several possibilities to overcome this problem.

Keywords: Phylogenetic inference problems
Polynomial-time algorithms

1 Introduction

Phylogenetic trees are commonly used to represent the evolutionary history of a set of taxa. The leaves represent extant taxa; internal nodes represent speciation events that caused lineages to diverge. If we assume the only processes

Research funded in part by the Netherlands Organization for Scientific Research (NWO), including Vidi grant 639.072.602, the 4TU Applied Mathematics Institute, the Natural Sciences and Engineering Research Council of Canada and the Canada Research Chairs program.

J. Jansson et al. (Eds.): AlCoB 2018, LNBI 10849, pp. 37–49, 2018.
https://doi.org/10.1007/978-3-319-91938-6_4

are speciation and modification and that no incomplete lineage sorting occurs, then any gene will give a gene tree that is consistent with the species phylogeny. In such cases, there exist efficient algorithms to reconstruct a species tree from gene trees. There are, however, evolutionary processes beyond vertical inheritance of genetic material and speciation events that make it more challenging to reconstruct the real evolutionary history. Examples of such processes are hybridization, horizontal gene transfer, and duplication. Each of these processes can result in discordance between gene trees.

This leads to a number of problems in which the task is to minimize the number of such complicating events. In *reconciliation problems*, we are given the gene trees together with the species phylogeny, and the task is to find optimal embeddings of the gene trees into the species phylogeny. Such methods are for example used to estimate dates of duplications, to discover relations between duplicate genes [7], and to reconstruct the infection history of parasites [19]. In *inference problems*, only the gene trees are given and we aim to find a species phylogeny that minimizes the discordance with the gene trees. Such problems are relevant when the species phylogeny is not yet known with certainty.

Duplication Minimization Problems. Gene duplications happen as a consequence of errors in the DNA replication process. This leads to a species having multiple copies of the same gene. There exist many types of gene duplication, which depend on the positions of errors within the replication process [20]. The scale of gene duplications is determined by the number of genes that get duplicated. An extreme example of a large-scale duplication is *Whole Genome Duplication (WGD)*, in which every gene in the genome is duplicated. This process, also known as polyploidization, occurs as a result of an error in separation of chromosomes during gamete production. It is most common in plants but has also occurred in animals [22], and there are two WGD events even in the evolutionary history leading to humans [8]. Large-scale duplications provide species with diversification potential, giving them the ability to quickly adapt to a changing environment [10].

In their seminal paper [11], Goodman et al. pioneered the parsimony approach to reconciling gene trees with species trees. This has motivated researchers to explore reconciliation through different models, whilst optimizing some measure of the number of duplication events.

These models can be categorized according to how duplication events are clustered to form duplication episodes and which restrictions are put on the possible locations of duplications [21]. We focus on the *minimal episodes (ME)* model where duplications can be clustered if they occur on the same branch of the species phylogeny and have no ancestor-descendant relationship in a gene tree (see Fig. 1). We believe this model to be most relevant since it can cluster duplications that can be part of a single (large-scale) duplication event. We consider the *unrestricted* variant of this model, which does not put any restrictions on the locations of gene duplications (called the FHS-model in [21]).

Reconciliation problems have been studied intensively, especially models without clustering. Several reconciliation problems with clustering have been

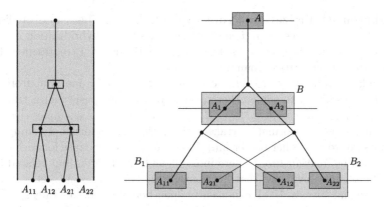

Fig. 1. Left: A gene tree embedded into a branch of a species tree with duplications clustered as in the Minimal Episodes model. Duplication clusters are shown as rectangles. Right: A representation of the DNA of the species at different points in the species tree (at corresponding heights). In the first duplication, the gene A (dark rectangle) is duplicated, forming A_1 and A_2. In the second duplication, the block B (light rectangle) comprising A_1 and A_2 is duplicated. This results in four homologous copies of gene A using only two duplication episodes. The gene tree is also drawn through the depictions of the DNA.

proven to be computationally intractable [9,17], whereas for others there are polynomial-time [3,6] or even linear-time [16,18,21] algorithms. For unrestricted ME reconciliation, there only exists an exponential-time algorithm [21], while the computational complexity of this problem is still unknown.

It has also been attempted to use reconciliation as a basis for inferring species phylogenies. For the unrestricted ME model, Burleigh et al. [5] used a brute-force approach on all possible species phylogenies. They observed that the unrestricted ME model fails to rank the true species tree among the top third of all topologies. It was suggested that a possible reason for this anomaly is that duplication episodes near the root are overly powerful under this model. A similar observation was made in a more recent reconciliation study [21]. However, neither article gives a mathematical explanation for this phenomenon. It should also be noted that, since the number of possible species phylogenies grows extremely quickly with the number of species, brute-force approaches are only feasible for very small data sets.

Inference problems are generally assumed to be computationally intractable. However, NP-hardness has been proven only for some restricted inference problem without clustering [17]. For an inference problem with restricted clustering (called gene duplication (GD) clustering in [21]), NP-hardness was suggested in [9] but not proven. Because of the suspected intractability of these problems, some heuristic inference approaches have been attempted using efficient algorithms for reconciliation (see, e.g., [12]).

Reticulation Minimization Problems. Another possible cause of discordance between gene trees is *reticulate evolution*, such as hybridization or horizontal gene transfer. In such cases, the evolutionary history is represented by a *phylogenetic network* rather than a tree.

Reticulate evolution can occur in nature when genetic material from one species is transmitted to some other species. In asexual species, such transfers are called *horizontal gene transfers (HGT)*. In bacteria, for example, this happens in nature by transformation (take-up from the environment) or conjugation (transmission from another bacterium). In sexual species, a cause for such transmissions can be *hybridization*, where individuals from different but related taxa mate. There is also evidence that horizontal gene transfers occur between multicellular sexual species. HGT can even happen between more distant species.

Gene trees that appear to be inconsistent may in fact simply take different paths through the network. This leads to a family of inference problems in which the aim is to find a phylogenetic network that is consistent with the gene trees and has the minimum number of *reticulation events* (nodes in the network with two ancestral branches). A phylogenetic network is often taken to be consistent with a gene tree if that tree is *displayed* by the network, which, roughly speaking, means that the gene tree can be drawn inside the network in such a way that each network branch contains at most one lineage of the gene tree. A more generous definition is to count a network as consistent with a gene tree if the tree is *weakly displayed* by the network [13,23]. Roughly speaking, this means that different lineages of the gene tree may "travel down" the same branch of the network, as long as any branching node in the tree coincides with a branching node in the network. In this case, the tree is also called a *parental tree* of the network. This models situations where a species has individuals carrying multiple homologous copies of a gene.

The HYBRIDIZATION NUMBER problem, in which we seek a network with the minimum number of reticulations displaying all input trees, has been well-studied. It has been shown that HYBRIDIZATION NUMBER is NP-hard already when the input consists of only two gene trees [4]. Furthermore, there are theoretical FPT algorithms for any fixed number of gene trees [15], but there are no practical algorithms that can handle instances with more than two input trees unless the number of taxa is extremely small.

In contrast, the PARENTAL HYBRIDIZATION problem, in which we seek a network with the minimum number of reticulations that weakly displays each input tree, was introduced only recently [23] and its computational complexity was open prior to this article. Our motivation for studying this problem is threefold:

(i) Since HYBRIDIZATION NUMBER is NP-hard, it is interesting whether relaxing the notion of a tree displayed by a network leads to an easier problem.

(ii) Since reticulation can lead to multiple homologous copies of a gene in a species, requiring that each gene tree is displayed by the network may lead us to overestimate the number of reticulations.

(iii) The problem of finding an optimal network that weakly displays a set of phylogenies arises as a crucial subproblem in a recent heuristic approach for constructing phylogenetic networks in the presence of hybridization and incomplete lineage sorting [23].

Structural Assumptions. We restrict to binary trees and networks. Unlike many papers in this area, we allow a network to contain *parallel arcs*, that is, pairs of arcs that join the same pair of nodes. Parallel arcs are normally omitted because, for most problems, it can either be shown that there exists an optimal solution without parallel arcs or it can be assumed that a realistic solution contains no parallel arcs. For example, any set of gene trees has an optimal hybridization network without parallel arcs. For the problems studied in this paper, however, an optimal solution may require parallel arcs. Considering this problem with the added restriction that parallel arcs are forbidden may be an interesting mathematical challenge; however, we do not believe it is biologically meaningful.

Explicit reasons to allow parallel arcs in networks are abundant. We give three: First, if one restricts a large network to a subset of the taxa, the natural restriction could have parallel arcs. Second, phylogenetic Markov models for character evolution behave differently if parallel arcs are suppressed. Third, polyploidization events often result from a sort of interspecific or intraspecific hybridization [2]; an intraspecific hybridization is most naturally represented by parallel arcs in the network.

Throughout this paper, we allow input trees to be multi-labeled, that is, each species may appear as a label of multiple leaves in a tree. This is natural for the problems we study, as gene duplication and reticulation can both lead to multiple homologous genes appearing in the genome of a single species.

Our Contributions. We show that both UNRESTRICTED MINIMAL EPISODES INFERENCE and PARENTAL HYBRIDIZATION reduce to the problem BEADED TREE, which we introduce in this paper. Using this reduction, we show that both problems can be solved in polynomial time by adapting Aho et al.'s classic algorithm for testing gene tree consistency [1]. Thereby, we provide the first polynomial-time algorithm for an inference problem with a duplication cluster model. Furthermore, we provide the first polynomial-time algorithm for constructing a phylogenetic *network* from gene trees.

We also show that optimal solutions to BEADED TREE have a restricted structure and this has corresponding implications for the optimal solutions to UNRESTRICTED MINIMAL EPISODES INFERENCE and PARENTAL HYBRIDIZATION that our algorithms produce. Moreover, we show that, in fact, *all* optimal solutions to UNRESTRICTED MINIMAL EPISODES INFERENCE have a restricted structure. Therefore, this model should be used with care. We end with a discussion of different ways to overcome these issues.

See [14] for the full version of this paper.

2   Preliminaries

We begin by defining *multi-labeled trees*, which form the input for all problems considered in this paper.

Definition 1. *Let X be a set of species. A* multi-labeled tree (MUL-tree) *on X is a directed acyclic graph with one node of in-degree 0 and out-degree 1 (the* root*) and with all other nodes having either in-degree 1 and out-degree 2 (*tree nodes*) or in-degree 1 and out-degree 0 (*leaf nodes or leaves*). Each leaf is labeled with an element of X. If each element of X labels at most one leaf, we call the MUL-tree a* tree.

Next, we define a *duplication tree*, which represents the evolutionary history of a set of species, including points at which duplication events occurred.

Definition 2. *Let X be a set of species. A* duplication tree *on X is a directed acyclic graph D with one node of in-degree 0 and out-degree 1 (the* root*), $|X|$ nodes of in-degree 1 and out-degree 0 (*leaf nodes or leaves*), and all other nodes having either in-degree 1 and out-degree 2 (*tree nodes*) or in-degree 1 and out-degree 1 (*duplication nodes*). The leaves are bijectively labeled with the elements of X. The* duplication number *of D is the number of duplication nodes it contains.*

Informally, a MUL-tree T is *consistent* with a duplication tree D if T can be drawn inside D so that branches of T duplicate only at duplication nodes of D, in the sense that both out-edges of a node of T may follow the same out-edge of the duplication node (see Fig. 2). We formalize this as follows:

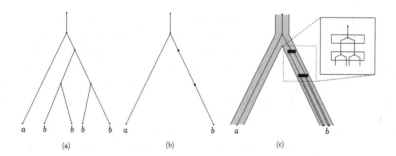

Fig. 2. (a) A MUL-tree T on $X = \{a, b\}$. (b) A duplication tree D that is consistent with T. (c) An illustration showing how T can be drawn inside D, and a zoomed-in portion to illustrate what happens at the duplication nodes. This shows how two or more incoming branches may duplicate simultaneously at a duplication node (according to the Minimal Episodes model).

Definition 3. *Given a MUL-tree T on X and a duplication tree D on X, a* duplication mapping *from T to D is a function $M : V(T) \to V(D)$ such that*

– *For each leaf $l \in L(T)$, $M(l)$ is a leaf of D labeled with the same species as l,*
– *For each edge $uv \in E(T)$, $M(u)$ is a strict ancestor of $M(v)$, and*
– *For each internal node u of T with children v, v', either $M(u)$ is the least common ancestor of $M(v)$ and $M(v')$, or $M(u)$ is a duplication node.*

This is illustrated in Fig. 3. We say that D is consistent with T if there is a duplication mapping from T to D.

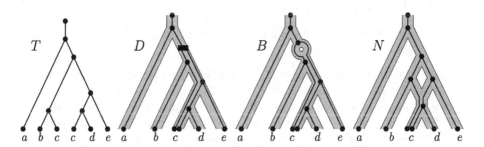

Fig. 3. A MUL-tree T, a duplication mapping from T to a duplication tree D, and weak embeddings of T into a beaded tree B and into a phylogenetic network N.

Let S be the species tree derived from D by suppressing duplication nodes. Then a duplication mapping from T to D represents a reconciliation of T with S under the Minimal Episodes model. Each duplication node in D represents a cluster of duplications, which is called a *duplication episode*. Internal nodes in T are treated as duplications if they are mapped to duplication nodes of D, and as speciations otherwise. Duplications are clustered together if they are mapped to the same duplication node of D. The properties of a duplication tree and duplication mapping ensure that duplications that are clustered occur on the same branch of the species phylogeny and have no ancestor-descendant relationship in a gene tree, as required by the Minimal Episodes model. We are now ready to define the following problem:

UNRESTRICTED MINIMAL EPISODES INFERENCE
Input: A set $\mathcal{T} = \{T_1, \ldots, T_t\}$ of MUL-trees with label sets $X_1, \ldots, X_t \subseteq X$.
Output: A duplication tree D on X with minimum duplication number such that D is consistent with each tree in $\mathcal{T}$.

Next, we introduce the concept of *phylogenetic networks*, which are central to the problem PARENTAL HYBRIDIZATION:

Definition 4. *Let X be a set of species. A (rooted binary) phylogenetic network N on X is a directed acyclic multigraph with one node of in-degree 0 and out-degree 1 (the root), $|X|$ nodes of in-degree 1 and out-degree 0 (leaves), and all other nodes having either in-degree 1 and out-degree 2 or in-degree 2 and out-degree 1 (reticulation nodes). The leaves are bijectively labeled with the elements of X. The* reticulation number *of N is the number of reticulation nodes it contains.*

Definition 5. *Given a set X of species, let N be a phylogenetic network, and T a MUL-tree on X. A weak embedding of T into N is a function h that maps every node of T to a node of N, and every edge in T to a directed path in N such that*

- *for each leaf $l \in L(T)$, $h(l)$ is a leaf of N labeled with the same species;*
- *for each edge $xy \in E(T)$, the path $h(xy)$ is a path from $h(x)$ to $h(y)$ in N;*
- *for each internal node x in T with children y, y', the paths $h(xy)$ and $h(xy')$ start with different out-edges of $h(x)$.*

This is illustrated in Fig. 3. We say that N weakly displays T if there is a weak embedding of T into N.

We note that N weakly displays T if and only if T is a *parental tree inside N* as defined in [23], hence the name PARENTAL HYBRIDIZATION. The notion of a tree *weakly displayed* by a network was first introduced in [13], where it was shown that T is weakly displayed by N if and only if there exists a *locally separated reconciliation* from T to N, which is equivalent to our definition of a weak embedding. We now formally define the PARENTAL HYBRIDIZATION problem:

PARENTAL HYBRIDIZATION
Input: A set $\mathcal{T} = \{T_1, \ldots, T_t\}$ of MUL-trees with label sets $X_1, \ldots, X_t \subseteq X$.
Output: A phylogenetic network N on X with minimum reticulation number such that N weakly displays all trees in $\mathcal{T}$.

Next, we define a certain type of phylogenetic network that, together with the corresponding computational problem defined below, turns out to be the key to both UNRESTRICTED MINIMAL EPISODES INFERENCE and PARENTAL HYBRIDIZATION.

Definition 6. *A bead in a phylogenetic network N is a pair of nodes (u, v) such that there are two parallel edges from u to v. A beaded tree is a phylogenetic network B in which every reticulation node is in a bead (see Fig. 3).*

BEADED TREE
Input: A set $\mathcal{T} = \{T_1, \ldots, T_t\}$ of MUL-trees with label sets $X_1, \ldots, X_t \subseteq X$.
Output: A beaded tree B on X with minimum reticulation number that weakly displays all trees in $\mathcal{T}$.

3   Reduction to Beaded Trees

The two problems UNRESTRICTED MINIMAL EPISODES INFERENCE and PARENTAL HYBRIDIZATION are in fact both reducible to BEADED TREE. This allows us to focus on the BEADED TREE problem in the rest of the paper.

Lemma 7. *Let X be a set of species and $\mathcal{T} = \{T_1, \ldots, T_t\}$ a set of MUL-trees on subsets of X. For any integer k, there exists a solution to UNRESTRICTED MINIMAL EPISODES INFERENCE on $\mathcal{T}$ with k duplications if and only if there exists a solution to BEADED TREE on $\mathcal{T}$ with k beads.*

Lemma 8. *For any instance $\mathcal{T}$ of* PARENTAL HYBRIDIZATION, *there exists an optimal solution B that is a beaded tree.*

We can also show that any instance of BEADED TREE has an optimal solution with a certain interesting structure.

Theorem 9. *Given an instance $\mathcal{T}$ of* BEADED TREE, *there exists an optimal solution B such that all reticulations are on a single path.*

Moreover, *any* optimal solution to an instance of BEADED TREE must satisfy certain structural properties.

Theorem 10. *Given any optimal solution B to an instance $\mathcal{T}$ of* BEADED TREE, *there exists a path from the root to a leaf of B, such that for any node u not on this path, there is at most one reticulation strictly descended from u.*

4 Beaded Tree Algorithm

Let SUPERTREE denote an algorithm that takes as input a set of MUL-trees $\mathcal{T}$, and returns either a tree T weakly displaying $\mathcal{T}$, or the value NONE if no such tree exists. A simple modification of the algorithm of [1] can be used for this.

Given a phylogenetic network N on X and a subset $S \subseteq X$, let $N \setminus S$ denote the network derived from N by deleting every leaf in S, and then exhaustively deleting unlabelled nodes of out-degree 0 and suppressing nodes of in-degree 1 out-degree 1. Let $N|_S$ denote the network $N \setminus (X \setminus S)$.

Given a set $\mathcal{T}$ of MUL-trees, let $F_1(\mathcal{T})$ denote the set of trees derived by, roughly speaking, splitting each tree of $\mathcal{T}$ into two by deleting the root.

Definition 11. *Let $\{T_1, \ldots, T_t\}$ be a set of MUL-trees and X the union of their label sets. The* split partition $\{S_1, \ldots, S_s\}$ *of $\{T_1, \ldots, T_t\}$ is the partition of X into minimal sets such that, if x and y appear within the same MUL-tree in $F_1(\mathcal{T})$ and $x \in S_j$, then $y \in S_j$.*

The beaded tree algorithm is described in Algorithm 1 and an example is given in Fig. 4.

Theorem 12. *Algorithm 1 finds an optimal solution to the* BEADED TREE *problem with input $\mathcal{T}$ in time $O((|X|^3 + |X|^2 k)n)$, with n the total number of vertices of the trees in $\mathcal{T}$ and k the reticulation number of an optimal solution.*

Data: $\mathcal{T} = \{T_1, \ldots, T_t\}$
Result: Beaded tree B that weakly displays $\mathcal{T}$ with minimum number of reticulations

if $|X| = 1$ *and* $\max_{i \in [t]} |L(T_i)| = 1$ **then**
   | **return** a tree with 1 leaf on X;
end
else
      | Calculate the split partition $S_1, \ldots, S_s$ of $\mathcal{T}$;
      | **for** $i \in [s]$ **do**
          | Let $T = \text{SUPERTREE}(\mathcal{T}|_{S_i})$;
          | **if** T *is not* NONE **then**
              | Let $B' = \text{BEADED-TREE}(\mathcal{T} \setminus S_i)$;
              | Construct B by joining B' and T with a new root;
              | **return** B
          | **end**
      | **end**
      | Let $B' = \text{BEADED-TREE}(F_1(\mathcal{T}))$;
      | Construct B by adding a bead whose child is the root of B';
      | **return** B
end

Algorithm 1. Algorithm BEADED-TREE($\mathcal{T}$).

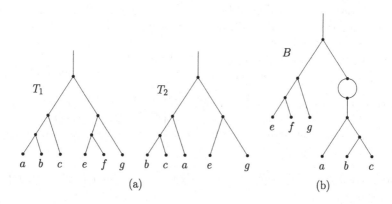

Fig. 4. (a) An instance $\mathcal{T} = \{T_1, T_2\}$ of BEADED TREE. (b) The beaded tree B constructed by running algorithm BEADED-TREE on $\mathcal{T}$. Initially, the split partition is $\{a, b, c\}, \{e, f, g\}$. As SUPERTREE returns a tree on $\{e, f, g\}$, the top tree node of B has that tree as one of its children. To construct the other side of B, we run BEADED-TREE on $\mathcal{T}|_{\{a,b,c\}}$, and SUPERTREE does not return a tree on this set. Therefore this side of B begins with a bead.

5   Concluding Remarks

Although we have shown that the UNRESTRICTED MINIMAL EPISODES INFERENCE and PARENTAL HYBRIDIZATION problems are polynomial-time solvable, we have also shown that the phylogenies produced by solving these problems have severely restricted structures.

The optimal phylogenetic network that our algorithm produces for the PARENTAL HYBRIDIZATION problem is always a phylogenetic tree with "beads", where a bead consists of a speciation directly followed by a reticulation. Such solutions are not necessarily the most realistic or likely ones since they contain a lot of "extra lineages", i.e. multiple lineages of an input tree travelling through the same branch of the phylogenetic network. Minimizing the total number of extra lineages, the *XL-score*, irrespective of the reticulation number, is also not ideal, since there always exists a solution with zero extra lineages and possibly a very high reticulation number. Therefore, the most relevant open problem that needs to be solved is to find a phylogenetic network that minimizes a weighted sum of the XL-score and the reticulation number of the network. Another alternative problem formulation that seems reasonable is to minimize the total number of parental trees that the constructed phylogenetic network has in addition to the input trees.

Another option would be to completely exclude beads in the solutions. However, although this is an interesting theoretical open problem, we do not see a reason why the resulting optimal solutions would by any more realistic, or why it would be reasonable to assume that a speciation cannot be followed by a reticulation.

Regarding UNRESTRICTED MINIMAL EPISODES INFERENCE, the situation is in some sense even worse. We have shown that *all* optimal solutions have a very specific structure: there is one main path from the root to a taxon containing potentially many duplication episodes, while each path branching off this main path contains at most one duplication episode. Although such scenarios are not to be excluded (for example see the eukaryotic species phylogeny from [12]), it is unrealistic to expect all phylogenies to look like this. Therefore, we have proposed an alternative problem in [14], which miminizes the "duplication depth": the maximum number of duplication episodes that lie on any directed path. This problem can also be solved in polynomial time and we expect it to produce more realistic solutions. Note moreover that, although the problem definition does not exclude unnecessary duplication episodes as long as they do not increase the duplication depth, our algorithm will not create such redundant duplication episodes. Nevertheless, to properly assess the two algorithms, it is necessary to implement both algorithms and extensively test them on simulated and real biological datasets.

Interestingly, the problem UNRESTRICTED MINIMAL EPISODES RECONCIL-IATION, where the species tree is given, is *not* known to be polynomial-time solvable. There is only an exponential-time algorithm [21]. Could it be possible to adapt our algorithm for UNRESTRICTED MINIMAL EPISODES INFERENCE to solve also the reconciliation variant?

Finally, it would be interesting to study more general models, which simultaneously take different processes into account, such as duplication episodes, hybridization, gene loss and transfers. Although such problems have been studied in a reconciliation setting where the species tree is (assumed to be) known, there has been less work on variants where the species tree or network needs to be inferred. Although such problems seem daunting, we have shown here that not knowing the species tree can actually make computational problems easier.

References

1. Aho, A.V., Sagiv, Y., Szymanski, T.G., Ullman, J.D.: Inferring a tree from lowest common ancestors with an application to the optimization of relational expressions. SIAM J. Comput. **10**, 405–421 (1981)
2. Albertin, W., Marullo, P.: Polyploidy in fungi: evolution after whole-genome duplication. Proc. Roy. Soci. Lond. B Biol. Sci. **279**(1738), 2497–2509 (2012)
3. Bansal, M.S., Eulenstein, O.: The multiple gene duplication problem revisited. Bioinformatics **24**(13), i132–i138 (2008)
4. Bordewich, M., Semple, C.: Computing the minimum number of hybridization events for a consistent evolutionary history. Discrete Appl. Math. **155**(8), 914–928 (2007)
5. Burleigh, J.G., Bansal, M.S., Eulenstein, O., Vision, T.J.: Inferring species trees from gene duplication episodes. In: Proceedings of the First ACM International Conference on Bioinformatics and Computational Biology, pp. 198–203. ACM (2010)
6. Burleigh, J.G., Bansal, M.S., Wehe, A., Eulenstein, O.: Locating multiple gene duplications through reconciled trees. In: Vingron, M., Wong, L. (eds.) RECOMB 2008. LNCS, vol. 4955, pp. 273–284. Springer, Heidelberg (2008). https://doi.org/10.1007/978-3-540-78839-3_24
7. Chan, Y.B., Ranwez, V., Scornavacca, C.: Reconciliation-based detection of co-evolving gene families. BMC Bioinform. **14**(1), 332 (2013)
8. Dehal, P., Boore, J.L.: Two rounds of whole genome duplication in the ancestral vertebrate. PLoS Biol. **3**(10), e314 (2005)
9. Fellows, M., Hallett, M., Stege, U.: On the multiple gene duplication problem. In: Chwa, K.-Y., Ibarra, O.H. (eds.) ISAAC 1998. LNCS, vol. 1533, pp. 348–357. Springer, Heidelberg (1998). https://doi.org/10.1007/3-540-49381-6_37
10. Glasauer, S.M., Neuhauss, S.C.: Whole-genome duplication in teleost fishes and its evolutionary consequences. Mol. Genet. Genomics **289**(6), 1045–1060 (2014)
11. Goodman, M., Czelusniak, J., Moore, G.W., Romero-Herrera, A.E., Matsuda, G.: Fitting the gene lineage into its species lineage, a parsimony strategy illustrated by cladograms constructed from globin sequences. Syst. Biol. **28**(2), 132–163 (1979)
12. Guigo, R., Muchnik, I., Smith, T.F.: Reconstruction of ancient molecular phylogeny. Mol. Phylogenet. Evol. **6**(2), 189–213 (1996)
13. Huber, K.T., Moulton, V., Steel, M., Wu, T.: Folding and unfolding phylogenetic trees and networks. J. Math. Biol. **73**(6–7), 1761–1780 (2016)
14. van Iersel, L., Janssen, R., Jones, M., Murakami, Y., Zeh, N.: Polynomial-time algorithms for phylogenetic inference problems (2018). arXiv:1802.00317 [q-bio.PE]
15. van Iersel, L., Kelk, S., Scornavacca, C.: Kernelizations for the hybridization number problem on multiple nonbinary trees. J. Comput. Syst. Sci. **82**(6), 1075–1089 (2016)
16. Luo, C.W., Chen, M.C., Chen, Y.C., Yang, R.W., Liu, H.F., Chao, K.M.: Linear-time algorithms for the multiple gene duplication problems. IEEE/ACM Trans. Comput. Biol. Bioinf. **8**(1), 260–265 (2011)

17. Ma, B., Li, M., Zhang, L.: From gene trees to species trees. SIAM J. Comput. **30**(3), 729–752 (2000)
18. Mettanant, V., Fakcharoenphol, J.: A linear-time algorithm for the multiple gene duplication problem. In: National Computer Science and Engineering Conference (Thailand) (2008)
19. Page, R.D.: Maps between trees and cladistic analysis of historical associations among genes, organisms, and areas. Syst. Biol. **43**(1), 58–77 (1994)
20. Panchy, N., Lehti-Shiu, M., Shiu, S.H.: Evolution of gene duplication in plants. Plant Physiol. **171**(4), 2294–2316 (2016)
21. Paszek, J., Gorecki, P.: Efficient algorithms for genomic duplication models. IEEE/ACM Trans. Comput. Biol. Bioinform. (2017)
22. Zhang, J.: Evolution by gene duplication: an update. Trends Ecol. Evol. **18**(6), 292–298 (2003)
23. Zhu, J., Yu, Y., Nakhleh, L.: In the light of deep coalescence: revisiting trees within networks. BMC Bioinform. **17**(14), 415 (2016)

Sequence Rearrangement and Analysis

Approximation Algorithms for Sorting Permutations by Fragmentation-Weighted Operations

Alexsandro Oliveira Alexandrino[1]([envelope]) [ID], Carla Negri Lintzmayer[2] [ID], and Zanoni Dias[1] [ID]

[1] Institute of Computing, University of Campinas (Unicamp),
Av. Albert Einstein 1251, Campinas, Brazil
`alexsandro.alexandrino@students.ic.unicamp.br`, `zanoni@ic.unicamp.br`
[2] Center for Mathematics, Computation and Cognition,
Federal University of ABC (UFABC), Av. dos Estados 5001, Santo André, Brazil
`carla.negri@ufabc.edu.br`

Abstract. Rearrangements are mutations that affect large portions of a genome. When comparing two genomes, one wants to find a sequence of rearrangements that transforms one into another. When we use permutations to represent the genomes, this reduces to the problem of sorting a permutation using some sequence of rearrangements. The traditional approach is to find a sequence of minimum length. However, some studies show that some rearrangements are more likely to disturb an individual, and so a weighted approach is closer to the real evolutionary process. Most weighted approaches consider that the cost of a rearrangement can be related to its type or to the number of elements affected by it. This work introduces a new type of cost function, which is related to the amount of fragmentation caused by a rearrangement. We present some results about lower and upper bounds for the fragmentation-weighted problems and the relation between the unweighted and the fragmentation-weighted approach. Our main results are 2-approximation algorithms for 5 versions of the fragmentation-weighted problem involving reversals and transpositions events.

Keywords: Genome rearrangements · Sorting permutations
Approximation algorithms

1 Introduction

One of the main problems in Computational Biology is to find the evolutionary distance among species. In most approaches, such distance only involves rearrangements, which are mutations that alter large pieces of the species' genome. Considering that the genome has no repeated genes, we can represent them as signed permutations, if the orientation of the genes is known, or unsigned permutations, if it is unknown. The most common types of rearrangements are reversals, which revert a segment of the genome, and transpositions, which exchange

© Springer International Publishing AG, part of Springer Nature 2018
J. Jansson et al. (Eds.): AlCoB 2018, LNBI 10849, pp. 53–64, 2018.
https://doi.org/10.1007/978-3-319-91938-6_5

two adjacent segments of the genome. A rearrangement model is the set of valid rearrangements that can be used to find the evolutionary distance, and it can contain one or more rearrangement operations.

The traditional approach is to consider that any rearrangement has the same probability to happen, and so the evolutionary distance can be defined as the minimum number of rearrangements that transforms one genome into another. Due to algebraic properties, this problem is equivalent to the problem of sorting permutations by rearrangements. The problems of sorting by reversals and sorting by transpositions are NP-Hard [5,6]. The best-known approximation factor for sorting by reversals is 1.375 [3], which is the same best-known factor for sorting by transpositions [8]. On the other hand, when considering signed permutations, the problem of sorting by reversals becomes polynomial, as shown by Hannenhalli and Pevzner [10].

Walter *et al.* [12] considered a variation in which both reversals and transpositions are allowed, and presented a 2-approximation algorithm for signed permutations and a 3-approximation algorithm for unsigned permutations. The best-known approximation factor for unsigned permutations, though, has factor $2k$ [11], where k is the approximation factor of the algorithm for cycle decomposition of the breakpoint graph [7]. The best-known value for k is $1.4167 + \epsilon$ [7], which turns the $2k$ into $2.8334 + \epsilon$, for $\epsilon > 0$. The complexity of these problems remains open.

Some rearrangements are more likely to occur than others [1,4], and so a weighted approach is closer to the real evolutionary process. In a weighted approach, each rearrangement has an associated cost and the goal is to find a minimum-cost sequence of rearrangements that transforms one genome into another, which is also equivalent to the sorting problem with weighted operations. The traditional approach is equivalent to setting a unitary cost for all rearrangements.

Eriksen [9] presented a study aimed to find which cost function for reversals and transpositions is more related to the evolutionary process. The best scenarios were found when the costs of reversals and transpositions are 2 and 3, respectively. The study did not consider prefix and suffix operations separately from other rearrangements.

In this work, we consider a new type of cost function which is equal to the amount of fragmentation that a rearrangement causes in a permutation. These costs are similar to those presented by Eriksen [9], differing only in prefix and suffix rearrangements, which cause less fragmentation in the genome. To our knowledge, there are no studies that have addressed this version of the problem.

We present 2-approximation algorithms for five such problems: Sorting by Fragmentation-Weighted Unsigned Reversals (SbR), Sorting by Fragmentation-Weighted Signed Reversals (SbR̄), Sorting by Fragmentation-Weighted Transpositions (SbT), Sorting by Fragmentation-Weighted Unsigned Reversals and Transpositions (SbRT), Sorting by Fragmentation-Weighted Signed Reversals and Transpositions (SbR̄T).

This work is organized as follows. Section 2 presents definitions related to the problems. Section 3 describes the algorithms we developed. Section 4 shows experimental results. Section 5 presents final considerations and future work.

2   Definitions

An *unsigned permutation* of size n is represented as $\pi = (\pi_1\ \pi_2\ \dots\ \pi_n)$, where $\pi_i \in \{1, 2, \dots, n\}$ and $\pi_i \neq \pi_j$ if and only if $i \neq j$, for all i and j. A *signed permutation* of size n is represented as $\pi = (\pi_1\ \pi_2\ \dots\ \pi_n)$, where $\pi_i \in \{-n, -(n-1), \dots, -1, 1, 2, \dots, n-1, n\}$ and $|\pi_i| \neq |\pi_j|$ if and only if $i \neq j$, for all i and j.

The *identity permutation* ι_n is equal to $(1\ 2\ \dots\ n)$. When considering signed permutations, all elements in the identity have positive sign. The *reverse permutation* η_n is the permutation $(n\ \dots\ 2\ 1)$, and the *signed reverse permutation* $\bar{\eta}_n$ is the permutation $(-n\ \dots\ -2\ -1)$.

A *composition* of two permutations π and σ is the operation "$\cdot$" for which $\pi \cdot \sigma = (\pi_{\sigma_1}\ \pi_{\sigma_2}\ \dots\ \pi_{\sigma_n})$. As we use permutations to represent rearrangements, the composition is used to indicate the occurrence of a rearrangement on a permutation. Thus, a rearrangement β transforms π into the permutation $\pi \cdot \beta$.

For all rearrangements β in the rearrangement model M, there is a cost associated to applying β in π, which is denoted by the function $f : M \to \mathbb{R}$. The cost of a sequence $\beta_1, \beta_2, \dots, \beta_m$ is equal to $\sum_{i=1}^{m} f(\beta_i)$.

Given a rearrangement model M, a cost function f, and a permutation π, the *sorting distance* $c_M^f(\pi)$ is the cost of a sequence $\beta_1, \beta_2, \dots, \beta_m$, such that all rearrangements are in M, $\pi \cdot \beta_1 \cdot \beta_2 \cdot \dots \cdot \beta_m = \iota_n$, and $\sum_{i=1}^{m} f(\beta_i)$ is minimum. When $f(\beta) = 1$ for all $\beta \in M$, this problem is equivalent to the traditional approach and is simply denoted by $c_M(\pi)$. Formally, the goal of sorting problems by fragmentation-weighted operations is to find such sorting distance.

2.1   Rearrangements

An *unsigned reversal* $\rho(i, j)$, with $1 \leq i < j \leq n$, is the rearrangement $(1\ 2\ \dots\ i-1\ j\ j-1\ \dots\ i+1\ i\ j+1\ \dots\ n)$ that, when applied to a permutation π, inverts the segment $\pi_i, \dots, \pi_j$, that is, it transforms π into $\pi \cdot \rho(i, j) = (\pi_1 \dots \pi_{i-1}\ \pi_j\ \pi_{j-1}\ \dots\ \pi_{i+1}\ \pi_i\ \pi_{j+1}\ \dots\ \pi_n)$. A *prefix reversal* $\rho_p(j)$ is a reversal $\rho(1, j)$, with $1 < j \leq n$, while a *suffix reversal* $\rho_s(i)$ is a reversal $\rho(i, n)$, with $1 \leq i < n$.

A *signed reversal* $\bar{\rho}(i, j)$, with $1 \leq i \leq j \leq n$, is the rearrangement $(1\ 2\ \dots\ i-1\ -j\ -(j-1)\ \dots\ -(i+1)\ -i\ j+1\ \dots\ n)$ that, when applied to a permutation π, inverts the segment $\pi_i, \dots, \pi_j$ and changes the sign of each element of the segment, that is, it transforms π into $\pi \cdot \bar{\rho}(i, j) = (\pi_1 \dots \pi_{i-1}\ -\pi_j\ -\pi_{j-1}\ \dots\ -\pi_{i+1}\ -\pi_i\ \pi_{j+1}\ \dots\ \pi_n)$. A *signed prefix reversal* $\bar{\rho}_p(j)$ is a reversal $\bar{\rho}(1, j)$, with $1 \leq j \leq n$, while a *signed suffix reversal* $\bar{\rho}_s(i)$ is a reversal $\bar{\rho}(i, n)$, with $1 \leq i \leq n$.

A *transposition* $\tau(i,j,k)$, with $1 \leq i < j < k \leq n+1$, is the rearrangement $(1\ 2\ \ldots\ i-1\ j\ j+1\ \ldots\ k-1\ i\ i+1\ \ldots\ j-1\ k\ \ldots\ n)$ that, when applied to a permutation π, exchanges the segment $\pi_i, \ldots, \pi_{j-1}$ with the adjacent segment $\pi_j, \ldots, \pi_{k-1}$, that is, it transforms π into $\pi \cdot \tau(i,j,k) = (\pi_1 \ldots \pi_{i-1}$ $\pi_j\ \pi_{j+1}\ \ldots\ \pi_{k-1}\ \pi_i\ \pi_{i+1}\ \ldots\ \pi_{j-1}\ \pi_k\ \ldots\ \pi_n)$. A *prefix transposition* $\tau_p(j,k)$ is a transposition $\tau(1,j,k)$, with $1 < j < k \leq n+1$, while a *suffix transposition* $\tau_s(i,j)$ is a transposition $\tau(i,j,n+1)$, with $1 \leq i < j \leq n$.

2.2 Breakpoints

A *reversal breakpoint* exists between a pair of consecutive elements π_i and π_{i+1} if $|\pi_{i+1} - \pi_i| \neq 1$, for $1 \leq i < n$. This type of breakpoint is used in problems involving unsigned reversals, such as SbR and SbRT. The number of reversal breakpoints in a permutation π is denoted by $b_r(\pi)$. Considering unsigned permutations, only the identity and reverse permutations have zero breakpoints.

A *transposition breakpoint* or *signed reversal breakpoint* exists between a pair of consecutive elements π_i and π_{i+1} if $\pi_{i+1} - \pi_i \neq 1$, for $1 \leq i < n$. This type of breakpoint is used in problems involving signed reversals or only transpositions, such as SbT, Sb$\bar{\text{R}}$, and Sb$\bar{\text{R}}$T. The number of transposition breakpoints or signed reversal breakpoints in a permutation π is denoted by $b_t(\pi)$ or $b_{\bar{r}}(\pi)$, respectively. For this type of breakpoint, only the identity and signed reverse permutations have zero breakpoints.

Considering some type of breakpoint, a *strip* is a maximal sequence of elements such that there are no breakpoints between two consecutive elements of the sequence. A *singleton* is a strip of length one. For unsigned permutations, a strip $(\pi_i\ \pi_{i+1}\ \ldots\ \pi_j)$ of length greater than one is *increasing* if $\pi_{k+1} = \pi_k + 1$ for all $i \leq k < j$, otherwise it is *decreasing*. A singleton is considered as an increasing strip. For signed permutations, we only differentiate *positive* and *negative* strips, where the former has only positive elements and the latter has only negative elements.

2.3 Cost Function

In this work, we consider that the cost of a rearrangement is equal to the amount of fragmentation that it causes in a permutation.

Formally, the cost function $f : M \rightarrow \mathbb{R}$, where M is the rearrangement model, for fragmentation-weighted reversals is defined as

$$f(\rho(i,j)) = \begin{cases} 0, & \text{if } i = 1 \text{ and } j = n \\ 1, & \text{if } i = 1 \text{ and } j < n \text{ or if } i > 1 \text{ and } j = n \\ 2, & \text{otherwise.} \end{cases}$$

For fragmentation-weighted transpositions, the cost function is defined as

$$f(\tau(i,j,k)) = \begin{cases} 1, & \text{if } i = 1 \text{ and } k = n+1 \\ 2, & \text{if } i = 1 \text{ and } k < n+1 \text{ or if } i > 1 \text{ and } k = n+1 \\ 3, & \text{otherwise.} \end{cases}$$

Henceforth, we will use r, $\bar{r}$, t, rt, and $\bar{r}t$ to denote the rearrangement model M that allows unsigned reversals, signed reversals, transpositions, unsigned reversals and transpositions, and signed reversals and transpositions, respectively.

3 Algorithms

The following subsections describe the approximation algorithms developed, in addition to some lower bounds on the distances of the problems.

3.1 Relation with Traditional Approach

This subsection shows how the traditional approach distance is related to the fragmentation-weighted distance, for any permutation. Lemma 1 shows a property of fragmentation-weighted problems involving reversals. Lemmas 2 to 4 define upper and lower bounds for the fragmentation-weighted distance considering the distance for the traditional approach (unweighted).

Lemma 1. *Considering the problems SbR, SbRT, Sb$\bar{R}$, and Sb$\bar{R}$T, there is an optimal sequence which contains at most one reversal $\rho(1, n)$ of cost 0.*

Proof. Any optimal sequence of length m that has $k > 1$ reversals of type $\rho(1, n)$ can be replaced by a sequence of length $m - k + (k \bmod 2)$ with at most one reversal of type $\rho(1, n)$, since any subsequence $\rho(1, n), \beta_1, \ldots, \beta_\ell, \rho(1, n)$, where β_i is a reversal or a transposition and $\ell \geq 0$, can be replaced by the subsequence $\beta'_1, \ldots, \beta'_\ell$, where β'_i is equal to $\rho(n + 1 - j, n + 1 - i)$, when $\beta_i = \rho(i, j)$, or $\tau(n + 1 - (k - 1), n + 1 - (j - 1), n + 1 - (i - 1))$, when $\beta_i = \tau(i, j, k)$.

Lemma 2. *For any permutation π, $c_r(\pi) \leq c_r^f(\pi) + 1$, $c_{rt}(\pi) \leq c_{rt}^f(\pi) + 1$, $c_{\bar{r}}(\pi)$ $\leq c_{\bar{r}}^f(\pi) + 1$, and $c_{\bar{r}t}(\pi) \leq c_{\bar{r}t}^f(\pi) + 1$.*

Proof. An optimal sorting sequence for a fragmentation-weighted problem is also a sorting sequence for the corresponding unweighted problem. By Lemma 1, we know that there exists an optimal sequence which contains at most one reversal of cost zero for the problems SbR, SbRT, Sb$\bar{R}$, and Sb$\bar{R}$T. Therefore, all rearrangements of this optimal sequence have cost greater than or equal to 1, except for at most one reversal, and so the number of rearrangements of this sequence is less than or equal to the cost of this sequence plus 1, for any π. Note that the unweighted distance, for any rearrangement model, is less than or equal to the number of rearrangements of any sequence that sorts π, which includes optimal sequences for the fragmentation weighted problems.

Lemma 3. *For any permutation π, $c_t(\pi) \leq c_t^f(\pi)$.*

Proof. This result follows from the fact that any transposition has cost greater than or equal to 1.

Lemma 4. *For any permutation π, $c_r^f(\pi) \leq 2c_r(\pi)$, $c_{\bar{r}}^f(\pi) \leq 2c_{\bar{r}}(\pi)$, and $c_t^f(\pi)$ $\leq 3c_t(\pi)$.*

Proof. First consider that only reversals are allowed. The maximum fragmentation cost that a reversal can have is 2, so the sequence from the unweighted problem, which can be used for the weighted problem, has cost less than or equal to twice its own length. Thus, any permutation can be sorted with a cost of at most $2c_r(\pi)$. The proof is analogous for signed reversals and for transpositions, noticing that the maximum fragmentation cost of a transposition is 3.

Lemmas 5 to 7 show approximation algorithms for SbR̄, SbR, and SbT.

Lemma 5. *SbR̄ is $2(1 + 1/(c_{\bar{r}}(\pi) - 1))$-approximable.*

Proof. There is an exact algorithm for sorting by unweighted signed reversals [10]. From Lemma 4, the fragmentation cost of the sequence retrieved by this algorithm is at most $2c_{\bar{r}}(\pi)$, for any permutation π. Therefore, by Lemma 2, the approximation factor for any permutation π, given by applying the fragmentation cost to the sequence retrieved by this algorithm, is at most $2c_{\bar{r}}(\pi)/(c_{\bar{r}}(\pi) - 1)$ $= 2(c_{\bar{r}}(\pi) - 1 + 1)/(c_{\bar{r}}(\pi) - 1) = 2(1 + 1/(c_{\bar{r}}(\pi) - 1))$ for SbR̄.

Lemma 6. *SbR is $2.75(1 + 1/(c_r(\pi) - 1))$-approximable.*

Proof. There is a 1.375-approximation algorithm for sorting by unweighted unsigned reversals [3]. From Lemma 4, the fragmentation cost of the sequence retrieved by this algorithm is at most $2 \times 1.375 \times c_r(\pi)$, for any permutation π. Therefore, by Lemma 2, the approximation factor for any permutation π, given by applying the fragmentation cost to the sequence retrieved by this algorithm, is at most $2 \times 1.375 \times c_r(\pi)/(c_r(\pi) - 1) = 2.75 \times (c_r(\pi) - 1 + 1)/(c_r(\pi) - 1)$ $= 2.75(1 + 1/(c_r(\pi) - 1))$ for SbR.

Lemma 7. *SbT is 4.125-approximable.*

Proof. There is a 1.375-approximation algorithm for sorting by unweighted transpositions [3]. From Lemma 4, the fragmentation cost of the sequence retrieved by this algorithm is at most $3 \times 1.375 \times c_t(\pi)$, for any permutation π. Therefore, by Lemma 3, the approximation factor for any permutation π, given by applying the fragmentation cost to the sequence retrieved by this algorithm, is at most $3 \times 1.375 \times c_t(\pi)/c_t(\pi) = 4.125$ for SbT.

3.2 Lower Bounds Using Breakpoints

In the next sections, we show how to achieve an approximation factor of 2 for all five problems. For that, we define new lower bounds using the concepts of breakpoints.

Lemma 8. *For any permutation π, $b_r(\pi) \leq c_r^f(\pi)$ and $b_{\bar{r}}(\pi) \leq c_{\bar{r}}^f(\pi)$.*

Proof. A reversal of cost 0 (inverts the whole permutation) does not remove any breakpoints. A prefix or suffix reversal has cost 1 and can remove at most one breakpoint. Any other reversal has cost 2 and removes at most two breakpoints. So, the ratio between cost and breakpoints removed is at least 1, and the cost of any sequence that sorts π must be greater than or equal to its number of breakpoints. The same is valid for signed reversals.

A similar result can be achieved for SbT problem, as shown in Lemma 9.

Lemma 9. *For any permutation π, $b_t(\pi) \leq c_t^f(\pi)$.*

Proof. A transposition of cost 1 can remove at most one breakpoint, since it exchanges the first segment of π with the last segment of π. A prefix or suffix transposition has cost 2 and can remove at most two breakpoints. A transposition of cost 3 removes at most three breakpoints. So, the cost of any sequence that sorts π must be at least the number of breakpoints.

3.3   2-Approximation Algorithms

The first 2-approximation algorithms that we present use the concepts of break-points and strips to achieve their approximation factor. The idea is to find the largest element that is not in the correct position, to bring the strip that contains this element to the correct place with one or two operations, and repeat this process until there are no breakpoints. At the end we can have the identity permutation (the goal of the sorting) or the reverse permutation, which can be transformed into the identity with a reversal of cost 0. As the next lemmas and theorems show, this approach results in 2-approximation algorithms for the five problems.

Lemma 10. *For any permutation π, $c_r^f(\pi) \leq 2b_r(\pi)$ and $c_{\bar{r}}^f(\pi) \leq 2b_{\bar{r}}(\pi)$.*

Proof. At each step, while the permutation is not sorted, do the following process. Let $|\pi_t| = k$ be the largest element (largest absolute value) out of its correct position (i.e., $k \neq t$ or $\pi_t < 0$) and let $\pi_i, \ldots, \pi_j$ be the strip containing k (note that $t = i$ or $t = j$). If it is an increasing (positive) strip, use one prefix reversal from positions 1 to j to put this strip at the beginning of π and another prefix reversal from 1 to k to put it in its correct place. Otherwise, the strip is decreasing (negative) and a reversal from position i to k can put it in its correct place. Note that for signed permutations, an element $\pi_i = -i$ is considered not to be in its correct position.

Note that if $\pi_n \neq n$, the first step of the algorithm will put the strip containing n (or $-n$) in the correct place with cost 1 but will not remove a breakpoint. All other steps remove a breakpoint, and they will use at most two operations of cost one for doing it. The last breakpoint removed, though, will cost at most one. Thus, $c_r^f(\pi) \leq 1 + 2(b_r(\pi) - 1) + 1 = 2b_r(\pi)$. Similarly, $c_{\bar{r}}^f(\pi) \leq 2b_{\bar{r}}(\pi)$.

Theorem 11. *SbR and SbR̄ are 2-approximable.*

Proof. This follows directly from Lemmas 8 and 10.

We call the approximation algorithms for SbR and SbR̄ given in Lemma 10 as 2-R and 2-R̄, respectively. Lemma 12 and Theorem 13 show that 2-R is also a 2-approximation algorithm for SbRT, and 2-R̄ is a 2-approximation algorithm for SbR̄T.

Lemma 12. *For any permutation* π, $c_{rt}^f(\pi) \leq 2b_r(\pi)$ *and* $c_{\bar{r}t}^f(\pi) \leq 2b_{\bar{r}}(\pi)$.

Proof. The proof follows directly from Lemma 10 and using the fact that $c_{rt}^f(\pi) \leq c_r^f(\pi)$ and $c_{\bar{r}t}^f(\pi) \leq c_{\bar{r}}^f(\pi)$.

Theorem 13. *SbRT and Sb$\bar{R}$T are 2-approximable.*

Proof. This follows directly from Lemmas 8, 9 and 12.

Next lemma describes the general idea behind algorithm 2-T, for SbT problem, and Theorem 15 proves its approximation factor.

Lemma 14. *For any unsigned permutation* π, $c_t^f(\pi) \leq 2b_t(\pi)$.

Proof. At each step, while the permutation is not sorted, perform the following process. Let $\pi_j = k$ be the largest element out of its correct position (i.e., $k \neq j$) and let $\pi_i, \ldots, \pi_j$ be the strip containing k. Move such strip to its correct place with a prefix transposition $\tau(1, j + 1, k + 1)$.

Note that if $\pi_n \neq n$, the first step of the algorithm will put the strip containing n in the correct place with cost 1 and it may not remove a breakpoint. All other steps remove a breakpoint with one operation, and they will use cost at most 2. The last breakpoint removed, though, will cost at most one, since the last transposition applied will remove the remaining two breakpoints in the permutation. Thus, $c_t^f(\pi) \leq 1 + 2(b_t(\pi) - 1) + 1 = 2b_t(\pi)$.

Theorem 15. *SbT is 2-approximable.*

Proof. This follows directly from Lemmas 9 and 14.

All three algorithms have time complexity $O(n^2)$, since the distance is $O(n)$ and they spend linear time to choose and to apply an operation at each step.

3.4 Greedy Algorithms

We also developed greedy algorithms, whose choice is to give priority to rearrangements which remove more breakpoints with the lowest cost. These algorithms work as follows. At each step, the rearrangement with best ratio "removed breakpoints/cost of operation" is chosen. This is repeated until there are no breakpoints left. For the problems SbR and Sb$\bar{R}$, we may reach a state where the permutation has only increasing (positive) strips or only decreasing (negative) strips. In this case, there are no operations that can remove breakpoints and so the algorithm uses a reversal of cost 1 to ensure that in the next step there will exist a rearrangement with ratio 1. At the end, when considering problems with reversals, we may have the reverse permutation, which can be sorted with a cost 0 reversal. These algorithms are named 2-Rg, 2-$\bar{R}$g, 2-Tg, 2-RTg, and 2-$\bar{R}$Tg for the problems SbR, Sb$\bar{R}$, SbT, SbRT, and Sb$\bar{R}$T, respectively. They are also 2-approximation algorithms as shown in Lemmas 16 to 18.

Lemma 16. *2-Rg and 2-R̄g are 2-approximation algorithms for SbR and SbR̄, respectively.*

Proof. We need to prove that in each step there is a way to remove a breakpoint with cost less than or equal to 2. The result will then follow by Lemma 8. For any unsigned permutation π, we can divide the proof into the following cases:

1. π *has at least one singleton.* Let π_i be the first singleton. If $\pi_i < n$, then there is a reversal that can remove a breakpoint by joining π_i with the strip containing $\pi_i + 1$. If $\pi_i > 1$, then there is a reversal that can remove a breakpoint by joining π_i with the strip containing $\pi_i - 1$.

2. π *has at least one strip of different type from the others.* Let $\pi_i, \ldots, \pi_j$ be the first strip such that there exists a strip $\pi_{i'}, \ldots, \pi_{j'}$ of different type for which $|\pi_j - \pi_{j'}| = 1$ or $|\pi_i - \pi_{i'}| = 1$. Note that we will always be able to find such strip if π respects this case. If $|\pi_j - \pi_{j'}| = 1$, a reversal $\rho(j+1, j')$ will join these strips and remove a breakpoint. Otherwise, $|\pi_i - \pi_{i'}| = 1$ and a reversal $\rho(i, i' - 1)$ will remove a breakpoint.

3. *All strips are increasing or all strips are decreasing.* In this case, there is no reversal that can remove a breakpoint, so the algorithm has to use a reversal of cost 1 to ensure that in the next step there will be a reversal that can remove a breakpoint with cost 1. Let $\pi_1, \ldots, \pi_j$ be the first strip. If $\pi_1, \ldots, \pi_j$ is increasing and $\pi_j < n$, a prefix reversal $\rho(1, j)$ turns π into a permutation that respects case 2, and so in the next step a prefix reversal $\rho(1, i' - 1)$, where $\pi_j + 1$ is in position i', can remove a breakpoint. If $\pi_1, \ldots, \pi_j$ is decreasing and $\pi_j > 1$, a prefix reversal $\rho(1, j)$ turns π into a permutation that respects case 2, and so in the next step a prefix reversal $\rho(1, i' - 1)$, where $\pi_j - 1$ is in position i', can remove a breakpoint. If $\pi_1, \ldots, \pi_j$ is increasing and $\pi_j = n$, a suffix reversal $\rho(i', n)$, where $\pi_{i'}$ is the first position of the strip s that contains $\pi_1 - 1$, makes the strip s be the last strip and a prefix reversal can join the first strip with s and remove a breakpoint. If $\pi_1, \ldots, \pi_j$ is decreasing and $\pi_j = 1$, the analysis is analogous to the previous one.

The proof is analogous for signed reversals. $\square$

Lemma 17. *2-Tg is a 2-approximation algorithm for SbT.*

Proof. At each step of 2-Tg, there will be at least one prefix or suffix transposition that removes one breakpoint in the following way. Let $\pi_1, \ldots, \pi_j$ be the first strip. If $\pi_j < n$, then there is a prefix or a suffix transposition that can join $\pi_1, \ldots, \pi_j$ with the strip containing $\pi_j + 1$, removing a breakpoint. Otherwise, we have $\pi_j = n$ and there is a prefix transposition than can join $\pi_1, \ldots, \pi_j$ with the strip containing $\pi_1 - 1$. So, the average cost of removing a breakpoint is less than or equal to 2. By Lemma 14, this is a 2-approximation algorithm. $\square$

Lemma 18. *2-RTg and 2-R̄Tg are 2-approximation algorithms for SbRT and SbR̄T, respectively.*

Proof. The proof is analogous to the proofs of Lemmas 16 and 17, since these algorithms use both reversals and transpositions. Note that in the only case where no reversals can remove a breakpoint, all strips are increasing or decreasing, and we can use a prefix or suffix transposition to remove one breakpoint with cost 2. The arguments are analogous to the ones used in Lemma 17.

The algorithms 2-Rg and 2-R̄g have time complexity $O(n^3)$, since they spend $O(n^2)$ time to choose the operation to be applied. The algorithms 2-Tg, 2-RTg, and 2-R̄Tg have time complexity $O(n^4)$, since they involve transpositions and need to spend $O(n^3)$ time to choose the operation to be applied.

4   Experimental Results

All algorithms were implemented in C language and executed on an Intel Core i7 with 8 cores of 4.20 GHz, 8 GB of RAM running Ubuntu 16.04 LTS. For the experimental results, we used two sets of permutations. One of them has 99,000 random unsigned permutations, being 1,000 of each size n, with n varying from 10 to 500 in intervals of 5. The other also has 99,000 permutations divided in the same way, but they are signed.

The algorithms used in this experiment were those presented in this work and five more algorithms from the literature for the unweighted case that were adapted to the fragmentation-weighted version. To adapt these algorithms, we took each sequence of operations returned by them and applied our fragmentation cost function. These algorithms are: A-R, $(1.4167 + \epsilon)$-approximation algorithm for sorting by unsigned reversals, which is an algorithm for cycle decomposition [7] followed by exact algorithm for signed reversals [10]; A-R̄, exact algorithm for sorting by signed reversals [10]; A-T, 1.5-approximation algorithm for sorting by transpositions [2]; A-RT, 3-approximation algorithm for sorting by unsigned reversals and transpositions [12]; A-R̄T, 2-approximation algorithm for sorting by signed reversals and transpositions [12].

The results are presented in Figs. 1(a) to (e), where the x-axis gives the size of the permutations and the y-axis gives the average approximation factor among all permutations of that size. The approximation factors were calculated using the theoretical lower bounds given in Lemmas 8 and 9.

As expected, the greedy algorithms had better practical results than the other algorithms. In all cases, except for the algorithm A-RT, the adapted algorithms had better results than the 2-approximation algorithms presented in Sect. 3.3. We can note that all greedy algorithms presented average approximation factors significantly smaller than the other algorithms.

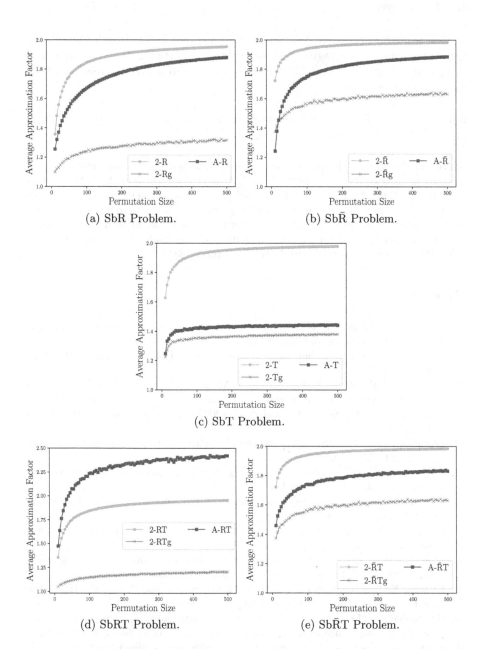

Fig. 1. Average approximation factor for 2-R, 2-Rg, A-R, 2-R̄, 2-R̄g, A-R̄, 2-T, 2-Tg, A-T, 2-RT, 2-RTg, A-RT, 2-R̄T, 2-R̄Tg, and A-R̄T when the permutation size grows. For all five problems, the greedy algorithms were significantly better than the others.

5   Conclusion

In this work we presented 2-approximation algorithms for 5 problems of sorting permutations by fragmentation-weighted operations. We also presented greedy algorithms and a relation between the unweighted approach and the fragmentation-weighted approach.

For future work, we aim at improving the approximation factors. For this, our next step will be the use of other structures, such as the cycle graph, to obtain better lower bounds and approximation factors.

Acknowledgments. This work was partially supported by the São Paulo Research Foundation, FAPESP (grants 2013/08293-7, 2015/11937-9, 2016/14132-4, 2017/12646-3, and 2017/16871-1), the National Counsel of Technological and Scientific Development, CNPq (grants 400487/2016-0, 425340/2016-3, and 131182/2017-0), and the program between the Brazilian Federal Agency for the Support and Evaluation of Graduate Education, CAPES, and the French Committee for the Evaluation of Academic and Scientific Cooperation with Brazil, COFECUB (grant 831/15).

References

1. Bader, M., Ohlebusch, E.: Sorting by weighted reversals, transpositions, and inverted transpositions. J. Comput. Biol. **14**(5), 615–636 (2007)
2. Bafna, V., Pevzner, P.A.: Sorting by transpositions. SIAM J. Discrete Math. **11**(2), 224–240 (1998)
3. Berman, P., Hannenhalli, S., Karpinski, M.: 1.375-Approximation algorithm for sorting by reversals. In: Möhring, R., Raman, R. (eds.) ESA 2002. LNCS, vol. 2461, pp. 200–210. Springer, Heidelberg (2002). https://doi.org/10.1007/3-540-45749-6_21
4. Blanchette, M., Kunisawa, T., Sankoff, D.: Parametric genome rearrangement. Gene **172**(1), GC11–GC17 (1996)
5. Bulteau, L., Fertin, G., Rusu, I.: Sorting by transpositions is difficult. SIAM J. Comput. **26**(3), 1148–1180 (2012)
6. Caprara, A.: Sorting permutations by reversals and eulerian cycle decompositions. SIAM J. Discrete Math. **12**(1), 91–110 (1999)
7. Chen, X.: On sorting unsigned permutations by double-cut-and-joins. J. Comb. Optim. **25**(3), 339–351 (2013)
8. Elias, I., Hartman, T.: A 1.375-approximation algorithm for sorting by transpositions. IEEE/ACM Trans. Comput. Biol. Bioinf. **3**(4), 369–379 (2006)
9. Eriksen, N.: Combinatorics of Genome Rearrangements and Phylogeny. Teknologie Licentiat Thesis, Kungliga Tekniska Högskolan, Stockholm (2001)
10. Hannenhalli, S., Pevzner, P.A.: Transforming cabbage into turnip: polynomial algorithm for sorting signed permutations by reversals. J. ACM **46**(1), 1–27 (1999)
11. Rahman, A., Shatabda, S., Hasan, M.: An approximation algorithm for sorting by reversals and transpositions. J. Discrete Algorithms **6**(3), 449–457 (2008)
12. Walter, M.E.M.T., Dias, Z., Meidanis, J.: Reversal and transposition distance of linear chromosomes. In: Proceedings of the 5th International Symposium on String Processing and Information Retrieval (SPIRE 1998), pp. 96–102. IEEE Computer Society, Los Alamitos (1998)

Heuristics for the Sorting
Signed Permutations by Reversals
and Transpositions Problem

Klairton Lima Brito[1]([⊠]) [ID], Andre Rodrigues Oliveira[1] [ID], Ulisses Dias[2] [ID],
and Zanoni Dias[1] [ID]

[1] Institute of Computing, University of Campinas,
Albert Einstein 1251, Campinas, Brazil
{klairton,andrero,zanoni}@ic.unicamp.br
[2] School of Technology, University of Campinas,
Paschoal Marmo 1888, Limeira, Brazil
ulisses@ft.unicamp.br

Abstract. We present two heuristics, *Sliding Window* and *Look Ahead*, to improve solutions for the Sorting Signed Permutations by Reversals and Transpositions Problem. To assess the heuristics, we implemented algorithms described in the literature to provide initial solutions. Despite the fact that we addressed a specific problem, both heuristics can be applied to many others within the area of genome rearrangement. When time is a crucial factor, *Sliding Window* is a better choice because it runs in linear time and improves the initial solutions in 76.4% of cases. If the quality of the solution is a priority, *Look Ahead* should be preferred because it improves the initial solutions in 97.6% of cases, but it runs in $\mathcal{O}(n^3 \times alg(n))$, where $alg(n)$ is the complexity of the algorithm given as input. By using these heuristics one may find a good tradeoff between running time and solution improvement.

Keywords: Genome rearrangement · Heuristics · Reversals
Transpositions

1 Introduction

Genome rearrangements affect large portions of the DNA sequence. They occur when chromosomes break at one or more locations and the pieces are reassembled in a different order. Due to the *Principle of Maximum Parsimony*, we approximate the evolutionary distance by the minimum number of events that transforms one genome into another. A Genome Rearrangement Problem aims at finding this minimum number, the so-called *rearrangement distance*.

Assuming no duplicated genes, we assign numbers to each gene to represent genomes as permutations of integers. If we know the relative orientation of the genes, we associate a sign (positive or negative) to each element of the permutation, resulting in a *signed permutation*; we omit this sign otherwise, resulting in an *unsigned permutation* (or simply *permutation*).

© Springer International Publishing AG, part of Springer Nature 2018
J. Jansson et al. (Eds.): AlCoB 2018, LNBI 10849, pp. 65–75, 2018.
https://doi.org/10.1007/978-3-319-91938-6_6

A reversal is a genome rearrangement event that inverts a segment of the genome, changing the order and the orientation of genes in this segment. A transposition swaps the position of two consecutive genome segments, keeping the order and the orientation of genes unchanged inside the segments.

To compute the distance between two genomes, we map one to the identity permutation defined as $\iota_n = (+1 \ldots +n)$ and use gene labels to map the other to an arbitrary permutation α. The goal is to transform α into ι—a sorting problem—using the minimum number of genome rearrangement events.

Hannenhalli and Pevzner [10] proved that the Sorting Signed Permutations by Reversals problem can be solved in polynomial time. Caprara [4] proved that the unsigned version is NP-hard. Bulteau and coauthors [3] proved that the Sorting Permutations by Transpositions problem is also NP-hard.

Sorting Signed Permutations by Reversals and Transpositions has unknown complexity, the same being true for the unsigned version. The best algorithm for the signed version has an approximation factor of 2 [14]. The best algorithm for the unsigned has an approximation factor of $2k$ [11], where k is the approximation factor of the algorithm used for cycle decomposition [5].

In this work, we present two heuristics to improve solutions from existing algorithms. Our heuristics produce smaller sorting sequences in the vast majority of cases when compared to those provided by the algorithms with no heuristics applied.

The paper is organized as follows. Section 2 presents notations and definitions. Section 3 details the heuristics. Section 4 shows the algorithms used to evaluate our heuristics. Section 5 reports the experiments. Section 6 concludes the manuscript.

2   Preliminaries

In genome rearrangement problems, we represent a genome as an n-tuple, where each element stands for a gene or blocks of genes. Assuming no duplicated genes, the n-tuple is a permutation $\pi = (\pi_1 \pi_2 \pi_3 \ldots \pi_n)$, where $\pi_i \in \{-n, -(n-1), \ldots, -2, -1, +1, +2, \ldots, +(n-1), +n\}$ such that $|\pi_i| \neq |\pi_j| \leftrightarrow i \neq j$. The positive or negative sign of an element indicates the orientation of the gene.

The composition between two permutations $\pi = (\pi_1 \ \pi_2 \ \ldots \ \pi_n)$ and $\sigma = (\sigma_1 \ \sigma_2 \ \ldots \ \sigma_n)$ results in a new permutation: $\alpha = \pi \circ \sigma = (\pi_{\sigma_1} \ \pi_{\sigma_2} \ \ldots \ \pi_{\sigma_n})$. If $\sigma_i < 0$, then $\alpha_i = -\pi_{|\sigma_i|}$, otherwise $\alpha_i = \pi_{\sigma_i}$.

The inverse of σ is a permutation σ^{-1} such that $\sigma \circ \sigma^{-1} = \iota_n$. The inverse σ^{-1} indicates the position and orientation in σ of each element i.

A reversal reverts the order of the segment $\{\pi_i, \pi_{i+1}, ..., \pi_j\}$ and also flips the signs of the elements. Therefore, a reversal $\rho(i, j)$ applied to π leads to $\pi \circ \rho(i, j) = (+\pi_1 \ \ldots \ +\pi_{i-1} \ \underline{-\pi_j \ \ldots \ -\pi_i} \ +\pi_{j+1} \ \ldots \ +\pi_n)$.

A transposition $\tau(i,j,k)$, $1 \leq i < j < k \leq n+1$, swaps the positions of two adjacent blocks. Therefore, a transposition $\tau(i,j,k)$ applied to π leads to $\pi \circ \tau(i,j,k) = (\pi_1 \ldots \pi_{i-1} \pi_j \ldots \pi_{k-1} \pi_i \ldots \pi_{j-1} \pi_k \ldots \pi_n)$.

The *distance* between π and σ, $d(\pi,\sigma)$, is the size of a minimum length sequence $\delta_1, \delta_2, \ldots, \delta_t$ of reversals and transpositions such that $\pi \circ \delta_1 \circ \delta_2 \ldots \delta_t = \sigma$. In this case, $d(\pi,\sigma) = t$.

Let $\iota_n = (+1 \ldots +n)$ be the identity permutation. A *sorting problem* is the distance between an arbitrary permutation $\alpha = (\alpha_1 \ldots \alpha_n)$ into ι_n. We denote the distance between α and ι_n by $d(\alpha, \iota_n) = d(\alpha)$.

The sorting problem may appear a particular case of rearrangement distance, but it has the same power of representation. Sorting α is equivalent to transforming π into σ if we consider $\alpha = \pi \circ \sigma^{-1}$. Note that $d(\pi,\sigma) = d(\pi \circ \sigma^{-1}, \sigma \circ \sigma^{-1}) = d(\alpha, \iota_n) = d(\alpha)$.

If we can sort α, we can also transform π into σ using the same sequence of operations. For example, let $\pi = (+6 \ +5 \ +1 \ +2 \ +4 \ +3)$ and $\sigma = (+2 \ -1 \ +4 \ -5 \ +3 \ +6)$, the inverse of σ is $\sigma^{-1} = (-2 \ +1 \ +5 \ +3 \ -4 \ +6)$. We compute $\alpha = \pi \circ \sigma^{-1} = (+6 \ -4 \ -2 \ +1 \ +3 \ +5)$. Applying a sorting sequence in α leads to $\alpha \circ \rho(2,4) \circ \tau(4,5,6) \circ \tau(1,2,7) \circ \rho(1,1) = \iota_6$. Applying the same operations in π leads to $\pi \circ \rho(2,4) \circ \tau(4,5,6) \circ \tau(1,2,7) \circ \rho(1,1) = \sigma$.

We obtain an *extended permutation* from π by inserting two new elements: $\pi_0 = +0$ and $\pi_{n+1} = n+1$. From now on, unless stated otherwise, permutations will be extended.

A *breakpoint* occurs in a pair π_i and π_{i+1} of π if $\pi_{i+1} - \pi_i \neq 1$, $0 \leq i \leq n$. We denote the number of breakpoints by $b(\pi)$. For $\pi = (+0 \cdot -2 \ -1 \cdot +4 \ +5 \cdot -3 \cdot +6)$, where "$\cdot$" represents a breakpoint, we have $b(\pi) = 4$. The identity permutation ι is the only with no breakpoints.

Breakpoints split a permutation into *strips*, which are maximal intervals without breakpoints. We do not add the elements π_0 and π_{n+1} to the scope of strips. For $\pi = (+0 \cdot -2 \ -1 \cdot +4 \ +5 \cdot -3 \cdot +6)$, we have three strips: $(-2 \ -1)$, $(+4 \ +5)$, and (-3).

Christie [6] created an algorithm to reduce a permutation π into a permutation $\pi_{reduced}$ such that $d(\pi) \leq d(\pi_{reduced})$. Four steps summarize the algorithm: (i) Remove the first strip if it starts with $+1$. (ii) Remove the last strip if it ends with $+n$. (iii) Replace each strip with the smallest element in it. (iv) Renumber the final sequence to obtain a valid permutation.

For example, let $\pi = (+1 \ +2 \ -9 \ -8 \ +5 \ +6 \ +7 \ +3 \ +4)$ be a permutation with four strips: $(+1 \ +2)$, $(-9 \ -8)$, $(+5 \ +6 \ +7)$, and $(+3 \ +4)$. We remove the first strip since it starts with $+1$, resulting in $(-9 \ -8)$, $(+5 \ +6 \ +7)$, and $(+3 \ +4)$. Then, we select the smallest element in each strip: $(-9 \ +5 \ +3)$. Finally, we renumber the final sequence to obtain the reduced permutation: $\pi_{reduced} = (-3 \ +2 \ +1)$.

3   Heuristics

We developed two heuristics, *Sliding Window* and *Look Ahead*, that extend previous approaches applied to unsigned permutations [7,8] and we assess them on the Sorting Signed Permutations by Reversals and Transpositions problem.

3.1   Sliding Window

Sliding Window uses a database that contains optimal sorting sequences for signed permutations of size up to nine [9]. It receives a permutation π and an algorithm alg as input and outputs a sequence of rearrangement events that sorts π.

The heuristic behaves as follows: we use alg to sort π and generate a sequence of permutations $S = [\pi^0, \ldots, \pi^z]$, such that $\pi^i \circ \delta = \pi^{i+1}$, where $\delta \in \{\rho, \tau\}$, for $0 \leq i < z$. The output is a sequence $S' = [\pi^0, \ldots, \pi^y]$, such that $y \leq z$ and $\pi^y = \pi^z$.

Initially, the heuristic picks a subsequence of permutations S^w from S that we call window. The window begins with π^i and ends with π^j, $0 \leq i < j \leq z$. The heuristic computes $\alpha = \pi^i \circ \pi^{j^{-1}}$ and reduces it to $\alpha_{reduced}$. If $\alpha_{reduced}$ has up to nine elements, we retrieve the optimal sorting sequence from the database, otherwise a smaller window S^w will be sought and slided through S.

If the optimal sequence for $\alpha_{reduced}$ is shorter than S^w, we use it to build a sequence $S^{w'}$ that sorts α. Each permutation $\alpha' \in S^{w'}$ is replaced by $\pi^j \circ \alpha'$ and the window S^w is replaced by $S^{w'}$, which improves S. Figure 1 shows a flowchart for this heuristic.

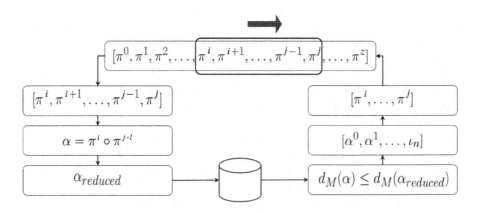

Fig. 1. Flowchart of the *Sliding Window* heuristic.

The heuristic runs in $\mathcal{O}(n + alg(n))$, where $alg(n)$ is the complexity of the algorithm given as input.

3.2   Look Ahead

Look Ahead receives a permutation π and an algorithm *alg* as input, and outputs a sequence of events that sorts π. The heuristic behaves as follows: we start with the permutation π as the current permutation. While the current permutation is not sorted, the heuristic assess all possible reversals and transpositions, fully investigating the neighborhood of π.

We use *alg* to estimate the distance of each permutation in the neighborhood of π, and we select the permutation with the shortest distance (or one of the shortest if multiple choices are available). The selected permutation will be the current permutation in the next iterative step. The process ends when we reach the identity.

Look Ahead requires a distance estimator *alg* to select an operation at each step. If the estimator does not work well, it negatively impacts the solution provided.

The heuristic runs in $\mathcal{O}(n^3 \times alg(n))$, where $alg(n)$ is the complexity of the algorithm given as input. Since the complexity of this heuristic is directly linked to $alg(n)$, it becomes prohibitive in cases where $alg(n)$ has a high complexity. Figure 2 shows the flowchart for this heuristic.

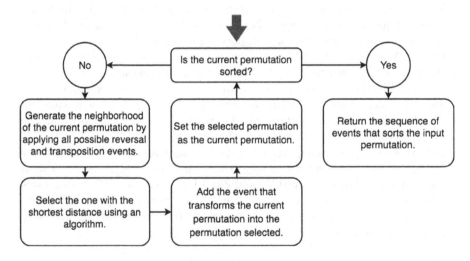

Fig. 2. Flowchart of the *Look Ahead* heuristic.

4   Algorithms Implemented to Evaluate the Heuristics

We use as input different algorithms from the literature. Some were not designed for the Sorting Signed Permutations by Reversions and Transpositions problem, but they provide a valid solution or some modifications were performed to make it valid. We employed such algorithms to verify the behavior on various situations. Table 1 shows the algorithms used as input.

Table 1. Algorithms used to evaluate the heuristics.

Rearrangement problem	Code	Reference	Time	Ratio
Reversal and transposition	RSH	Rahman *et al.* [11]	$\mathcal{O}(n^3)$	$2k$
Signed reversal and transposition	WDM	Walter *et al.* [14]	$\mathcal{O}(n^3)$	2
	BRPT	Walter *et al.* [14]	$\mathcal{O}(n^2)$	3
	BRPR	Walter *et al.* [14]	$\mathcal{O}(n^2)$	3
Signed reversal	HPB	Hannenhalli and Pevzner [10]	$\mathcal{O}(n^2)$	1
		Bader *et al.* [1]	$\mathcal{O}(n)$	1
Transposition	BP	Bafna and Pevzner [2]	$\mathcal{O}(n^2)$	1.5

- RSH: An algorithm for the Sorting Unsigned Permutations by Reversals and Transpositions problem with an approximation factor of $2k$, where k is the approximation of the algorithm that decomposes π in cycles. If applied on signed permutations, it outputs valid solution with approximation factor of 2.
- WDM: An algorithm for the Sorting Signed Permutations by Reversals and Transpositions problem that guarantees an approximation factor of 2.
- BRPT: An algorithm for the Sorting Signed Permutations by Reversals and Transpositions problem with an approximation factor of 3. The algorithm greedily removes the largest number of breakpoints. In case of ties between reversals and transpositions, a transposition is chosen.
- BRPR: A variation of BRPT that favours reversals instead of transpositions.
- HPB: An exact algorithm for the Sorting Signed Permutations by Reversals problem. Since *Look Ahead* needs a distance estimation, we used a linear time algorithm that outputs only the distance. Since *Sliding Window* requires an initial sequence of rearrangement events, we used a quadratic algorithm. The implementations were provided by Tesler [12,13].
- BP: An approximation algorithm for the Sorting Unsigned Permutations by Transpositions problem with an approximation factor of $\frac{3}{2}$. To ensure a valid solution for the Sorting Signed Permutations by Reversals and Transpositions problem, we first reverse all negative strips before applying this algorithm. The final sorting sequence is composed by the reversal operations that were first applied and the result of this algorithm.

5   Results

The heuristics and the algorithms implemented from literature received the same set of permutations that were randomly generated with the maximum number of breakpoints. The sizes of permutations ranged from 10 to 500 and increased in intervals of 10 from 10 to 100, and in intervals of 50 from 150 up to 500. For each size, we created a set of 1000 permutations. We executed *Look Ahead* on permutations with size up to 100 due to the slow running time. We executed *Sliding Window* on all permutations.

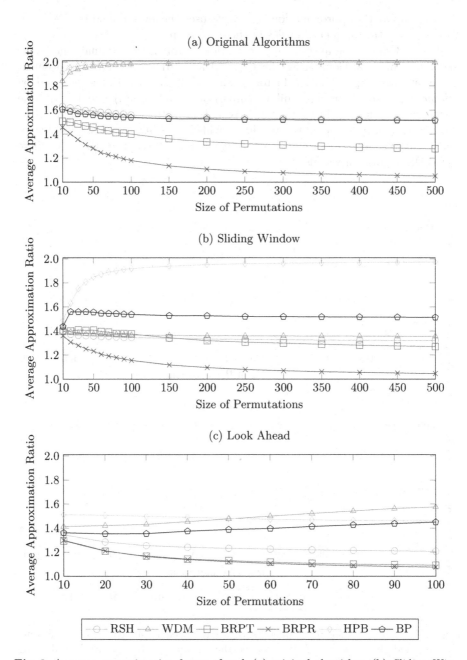

Fig. 3. Average approximation factor of each (a) original algorithm, (b) *Sliding Window*, and (c) *Look Ahead*. We can see a significant improvement in the average approximation factor in almost all the algorithms where *Sliding Window* was applied. *Look Ahead* improved the average approximation factor of all algorithms.

To compute the approximation factors, we used the lower bound $\lceil \frac{(n+1)-c(\pi)}{2} \rceil$, where $c(\pi)$ is the number of cycles in the cycle graph [14].

Figure 3 shows the average approximation factor of the original algorithms and our heuristics. Comparing the Fig. 3(a) and (b) we observe improvement in the average approximation factor in almost all the algorithms provided by *Sliding Window*. We make similar comparison with Fig. 3(a) and (c) and see a significant reduction in average approximation factor in all algorithms using *Look Ahead*. In most cases, the results provided by *Look Ahead* showed better performance than those provided by *Sliding Window*, except for the case where the WDM algorithm was used.

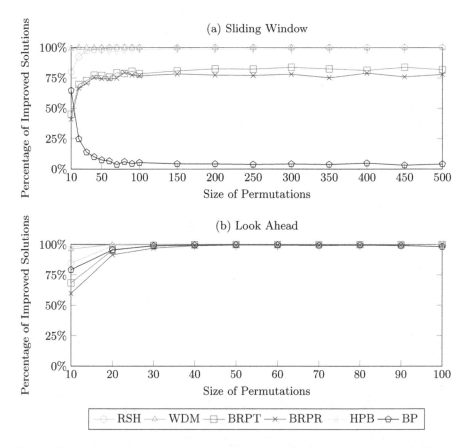

Fig. 4. Percentage of sorting sequences that have been improved using (a) *Sliding Window* and (b) *Look Ahead*. We can see that in almost all cases *Sliding Window* improved the initial sorting sequence of a significant amount of permutations. The only case in which this behavior was not observed was when we used the BP algorithm. *Look Ahead* improved the initial sorting sequence of a significant amount of permutations. For all permutations with size greater than 10, this value exceeded 90%.

Figure 4 shows the percentage of permutations where the original sorting sequence was improved by Sliding Window and *Look Ahead*, respectively.

Applying *Sliding Window* in permutations of size 10 up to 100, we obtained an improvement in 75.6% of cases, and for permutations of size 150 up to 500, we obtained an improvement in 77.4% of cases. *Look Ahead* was executed with permutations up to size 100 and improved the original sorting sequence in 97.6% of cases. All algorithms presented significant improvements.

Table 2 reports the average running time in seconds. The abbreviations ALG, SW, LA, represents the original algorithm, *Sliding Window*, and *Look Ahead*. We see that *Sliding Window* runs extremely fast, whereas *Look Ahead* is more time-consuming.

Table 2. Average running time in seconds. In all cases, *Sliding Window* outputs a solution in less than 0.1 s. *Look Ahead* is more time-consuming, but it runs fast when an algorithm with low time complexity like HBP is used.

Algorithm	Size of permutations				
	100			500	
	ALG	SW	LA	ALG	SW
RSH	0.003	0.016	11378.694	0.048	0.085
WDM	0.003	0.017	15330.229	0.047	0.092
BRPT	0.001	0.007	1676.322	0.006	0.021
BRPR	0.001	0.007	1445.922	0.005	0.019
HPB	0.003	0.009	72.803	0.029	0.049
BP	0.002	0.005	3941.291	0.025	0.035

Table 3. Average approximation factor provided by the original algorithms and our heuristics. *Look Ahead* significantly improved the average approximation factor of all algorithms. *Sliding Window* showed better results when applied to specific algorithms for the Sorting Signed Permutations by Reversals and Transpositions Problem.

Algorithm	Permutation size				
	100			500	
	ALG	SW	LA	ALG	SW
RSH	1.559	1.347	1.208	1.520	1.322
WDM	1.982	1.368	1.575	1.997	1.356
BRPT	1.402	1.376	1.091	1.277	1.272
BRPR	1.180	1.155	1.075	1.052	1.046
HPB	1.990	1.920	1.459	1.998	1.974
BP	1.540	1.539	1.450	1.514	1.514

Table 3 shows a comparison between the average approximation factor of the original algorithms and our heuristics. The abbreviations ALG, SW, LA, represents the original algorithm, *Sliding Window*, and *Look Ahead*.

6   Conclusion

The heuristics presented in this work significantly improved the sorting sequence provided by several algorithms known in the literature for the Sorting Signed Permutations by Reversals and Transpositions Problem. The heuristics *Sliding Window* and *Look Ahead* improved the sorting sequence in 76.4% and 97.6% of cases, respectively.

These heuristics can be applied in scenarios with different needs. If time is a crucial factor, the *Sliding Window* stands out since it presents good results and suffer less variation in execution time when permutation size increases. If time is not a priority, then *Look Ahead* is a better fit, presenting more remarkable results.

The next step is to use these heuristics on variants of the Sorting Signed Permutations by Reversals and Transpositions problem and check if it is possible to obtain results similar to those shown in this work.

Acknowledgments. The authors acknowledge the support from CAPES, the International Cooperation Program CAPES/COFECUB Foundation under grant 831/15, the CNPq under grants 425340/2016-3, 400487/2016-0, 140466/2018-5, and 138219/2016-8 and also the São Paulo Research Foundation (FAPESP) under grants 2013/08293-7, 2014/19401-8, and 2015/11937-9.

References

1. Bader, D.A., Moret, B.M.E., Yan, M.: A linear-time algorithm for computing inversion distance between signed permutations with an experimental study. J. Comput. Biol. **8**, 483–491 (2001)
2. Bafna, V., Pevzner, P.A.: Sorting by transpositions. SIAM J. Discrete Math. **11**(2), 224–240 (1998)
3. Bulteau, L., Fertin, G., Rusu, I.: Sorting by transpositions is difficult. SIAM J. Comput. **26**(3), 1148–1180 (2012)
4. Caprara, A.: Sorting permutations by reversals and eulerian cycle decompositions. SIAM J. Discrete Math. **12**(1), 91–110 (1999)
5. Chen, X.: On sorting unsigned permutations by double-cut-and-joins. J. Comb. Optim. **25**(3), 339–351 (2013)
6. Christie, D.A.: Genome rearrangement problems. Ph.D. thesis, Department of Computing Science, University of Glasgow (1998)
7. Dias, U., Dias, Z.: Extending Bafna-Pevzner algorithm. In: Proceedings of the 1st International Symposium on Biocomputing (ISB 2010), pp. 1–8. ACM, New York (2010)
8. Dias, U., Galvão, G.R., Lintzmayer, C.N., Dias, Z.: A general heuristic for genome rearrangement problems. J. Bioinform. Comput. Biol. **12**(3), 26 (2014)

9. Galvão, G.R., Dias, Z.: An audit tool for genome rearrangement algorithms. J. Exp. Algorithmics **19**, 1–34 (2014)

10. Hannenhalli, S., Pevzner, P.A.: Transforming cabbage into turnip: polynomial algorithm for sorting signed permutations by reversals. J. ACM **46**(1), 1–27 (1999)

11. Rahman, A., Shatabda, S., Hasan, M.: An approximation algorithm for sorting by reversals and transpositions. J. Discrete Algorithms **6**(3), 449–457 (2008)

12. Tesler, G.: Efficient algorithms for multichromosomal genome rearrangements. J. Comput. Syst. Sci. **65**(3), 587–609 (2002)

13. Tesler, G.: GRIMM: genome rearrangements web server. Bioinformatics **18**(3), 492–493 (2002)

14. Walter, M.E.M.T., Dias, Z., Meidanis, J.: Reversal and transposition distance of linear chromosomes. In: Proceedings of the 5th International Symposium on String Processing and Information Retrieval (SPIRE 1998), Santa Cruz de La Sierra, Bolivia, pp. 96–102. IEEE Computer Society (1998)

Sorting Permutations
by Limited-Size Operations

Guilherme Henrique Santos Miranda[1]([⊠]) [iD], Carla Negri Lintzmayer[2] [iD],
and Zanoni Dias[1] [iD]

[1] Institute of Computing, University of Campinas (Unicamp),
Av. Albert Einstein 1251, Campinas, Brazil
`guilherme.miranda@students.ic.unicamp.br`, `zanoni@ic.unicamp.br`
[2] Center for Mathematics, Computation and Cognition,
Federal University of ABC (UFABC), Av. dos Estados 5001, Santo André, Brazil
`carla.negri@ufabc.edu.br`

Abstract. Estimating the evolutionary distance between genomes of
two organisms is a challenging task for Computational Biology. One of
the most well-accepted ways to do this is to consider the size of the smallest sequence of rearrangement events required to transform one genome
into another, characterizing the rearrangement distance problem. Computationally, genomes can be represented as permutations of integers
and, with this, the problem can be reduced to transforming a permutation into the identity with the minimum number of operations (sorting
the permutation). These operations are given by a rearrangement model
and they affect segments of a genome in different ways. Among the most
common models are those that allow only reversals, only transpositions,
or both of them. In this paper we study sorting permutations when a
restriction of biological relevance is added: the size of the rearrangements
should be at most a given value λ. Some results are known for $\lambda = 2$ and
$\lambda = 3$ but, to the best of our knowledge, there are no results for $\lambda > 3$.
We consider rearrangement models that allow reversals and/or transpositions for sorting unsigned permutations given any value of λ. We present
approximation algorithms for 3 such problems, where the approximation
factors depend on λ and/or on the size of the permutations.

Keywords: Genome rearrangement · Sorting permutations
Reversals · Transpositions · Computational Biology

1 Introduction

An important challenge to Computational Biology is to understand the evolutionary process of two organisms, which can be done by using the length of the
shortest sequence of genome rearrangements that transform one genome into the
other, called rearrangement distance. This is the most likely to occur, based the
Principle of Maximum Parsimony. A genome rearrangement is an event which
occurs with relative rarity, changing larger stretches of the genome.

© Springer International Publishing AG, part of Springer Nature 2018
J. Jansson et al. (Eds.): AlCoB 2018, LNBI 10849, pp. 76–87, 2018.
https://doi.org/10.1007/978-3-319-91938-6_7

Computationally, a genome can be represented by a permutation of integers if we assume that there are no duplicate genes and that its composition is a single linear chromosome. The permutation is called *unsigned permutation* when the orientation of genes are unknown. Due to such representation, estimating the evolutionary distance with the rearrangement distance can be reduced to calculating the minimum number of rearrangements that transform a permutation into another. In addition, one of the genomes can be represented as the identity permutation and, in this way, the problem is equivalent to finding the minimum number of operations that sort a permutation. For convenience, we will refer to *unsigned* permutations only as permutations in the rest of the text.

Different rearrangement events are considered in the literature. Two of the most often studied ones are (i) reversals, which invert a determined segment of the genome, and (ii) transpositions, which exchange two adjacent segments in the genome. The problem of sorting permutations by reversals was proved to be NP-Hard [4] and the best-known result is a 1.375-approximation algorithm, proposed by Berman *et al.* [2]. The problem of sorting permutations by transpositions was introduced by Bafna and Pevzner [1]. This problem was proved to be NP-Hard by Bulteau *et al.* [3] and the best known result is a 1.375-approximation algorithm which was proposed by Elias and Hartman [7].

The complexity of the problem of sorting permutations when both operations are allowed is unknown. Walter *et al.* [17] presented a 3-approximation algorithm for this problem and, later, Rahman *et al.* [15] presented a 2α-approximation algorithm, where α is the approximation factor of the cycle decomposition algorithm for breakpoint graphs [13]. Given the best-known value of α [13], the approximation factor is $2.8334 + \epsilon$, where $\epsilon > 0$.

These variants of sorting permutations problems can have extra restrictions, such as in which parts of the genome the operations will be applied [6,14] and the size limit of the rearrangements [5,8–11,16], which is the amount of permutation's elements that are affected by it. The biological relevance for considering limited-size operations is based on the observation that rearrangement events which modify large stretches of the genome occurs rarely, prevailing rearrangements that involve few genes [12]. Chen and Skiena [5] considered sorting permutations by applying only reversals of the same size and they characterized the number of equivalence classes of permutations of size n and reversals of size k, for linear and circular permutations. When the size limit of an operation is at most 2, Jerrum [10] proved that the problem of sorting permutations by reversals (or transpositions) is polynomial. When the limit is at most 3, the best results are (i) a 2-approximation [9] algorithm for the problem of sorting permutations by reversals, (ii) a $\frac{5}{4}$-approximation algorithm for the problem of sorting permutations by transpositions [11], and (iii) a 2-approximation [16] algorithm for the problem of sorting permutations by reversals and transpositions, respectively.

We present approximation algorithms for the problems of sorting permutations by reversals, by transpositions, and by both of them with an additional restriction: the operations have size limited by a given value λ. To the best of our knowledge, there are no results in the literature for $\lambda > 3$.

The next sections are organized as follows. Section 2 presents the definitions and notations regarding problems of sorting permutations. Section 3 presents the proposed approximation algorithms. Section 4 presents the concluding remarks.

2 Definitions

A *permutation* of size n is defined as $\pi = (\pi_1 \pi_2 \ldots \pi_n)$, where $\pi_i \in \{1, 2, \ldots, (n-1), n\}$ and $\pi_i \neq \pi_j$ if and only if $i \neq j$. The *identity permutation* of size n is defined as $\iota = (1 2 \ldots n)$.

Given two permutations π and σ of size n, the composition operation "." is defined as $\pi \cdot \sigma = (\pi_{\sigma_1} \pi_{\sigma_2} \ldots \pi_{\sigma_n})$. Compositions are used to indicate the application of a rearrangement over a permutation, as we see in the following.

A *reversal* $\rho(i, j)$ with $1 \leq i < j \leq n$ is an event that occurs in a permutation $\pi = (\pi_1 \pi_2 \ldots \pi_n)$ and transforms it into the permutation $\pi \cdot \rho(i, j) = (\pi_1 \pi_2 \ldots \pi_{i-1} \pi_j \pi_{j-1} \ldots \pi_{i+1} \pi_i \pi_{j+1} \ldots \pi_{n-1} \pi_n)$. For example, for the permutation $\pi = (1 2 3 4 5)$, we have $\pi \cdot \rho(2, 4) = (1 \underline{4 3 2} 5)$.

A *transposition* $\tau(i, j, k)$ with $1 \leq i < j < k \leq n + 1$ is an operation that occurs in a permutation $\pi = (\pi_1 \pi_2 \ldots \pi_n)$ and transforms it into the permutation $\pi \cdot \tau(i, j, k) = (\pi_1 \pi_2 \ldots \pi_{i-1} \pi_j \ldots \pi_{k-1} \pi_i \ldots \pi_{j-1} \pi_k \ldots \pi_n)$. For example, for the permutation $\pi = (1 2 3 4 5)$ we have $\pi \cdot \tau(1, 3, 5) = (\underline{3 4 1 2} 5)$.

A λ-*reversal* is a reversal $\rho(i, j)$ such that $j - i + 1 \leq \lambda$, where $j - i + 1$ is the *size* of the reversal. A λ-*transposition* is a transposition $\tau(i, j, k)$ such that $k - i \leq \lambda$, where $k - i$ is the *size* of the transposition.

In a sorting problem, we have a *rearrangement model*, denoted by β, which indicates what are the operations allowed to be applied in order to sort a permutation. This is used in the definitions bellow.

Given a rearrangement model β and a permutation π, the *sorting distance*, denoted by $d_\beta(\pi)$, is the minimum amount of operations in β needed to transform π into ι. If β allows only reversals, only transpositions, or both these operations, we denote the sorting distance of π by $d_r(\pi), d_t(\pi)$, and $d_{rt}(\pi)$, respectively. Similarly, we denote by $d_r^\lambda(\pi)$, $d_t^\lambda(\pi)$ and $d_{rt}^\lambda(\pi)$ the sorting distances for when we only allow λ-operations.

3 Approximation Algorithms

In this section we present approximation algorithms for the three problems we are studying. We separated this section in two parts. The first one presents algorithms whose approximation factors are better for large values of λ, while the second one presents algorithms whose approximation factors are better for small values of λ. We note, however, that all algorithms presented in both parts work for any $\lambda \geq 2$.

3.1   Approximation Algorithms for Large Values of λ

The approximation algorithms shown in this part were obtained by using algorithms that already exist in the literature for the variants where the size of the rearrangements is not limited. Thus, the first step is to relate the distance of our problems with the distance of such variants, which is shown in Lemma 1.

Lemma 1. *For all permutations π and all $\lambda \geq 2$, we have $d_r^\lambda(\pi) \geq d_r(\pi)$, $d_t^\lambda(\pi) \geq d_t(\pi)$ and $d_{rt}^\lambda(\pi) \geq d_{rt}(\pi)$.*

Proof. Any sorting sequence where the size of the rearrangements is limited by λ is valid for the case with no restriction.   □

Lemmas 2 and 3 show how to mimic any given reversal and transposition with a sequence of λ-reversals and λ-transpositions, respectively.

Lemma 2. *For a permutation π and $\lambda \geq 2$, the effect of a reversal $\rho(i,j)$ of size $j - i + 1 > \lambda$ can be obtained by at most $\frac{q(q+1)}{2}$ λ-reversals, where $q = \left\lceil \frac{j-i+1}{\lfloor \lambda/2 \rfloor} \right\rceil$.*

Proof. Initially we divide the segment of π from position i to position j into subsegments of size $\lfloor \lambda/2 \rfloor$, except maybe for the one closest to j, which results in $q = \left\lceil \frac{j-i+1}{\lfloor \lambda/2 \rfloor} \right\rceil$ subsegments. Formally, for each $1 \leq \ell < q$, the ℓth subsegment contains elements of π from position $i + \lfloor \lambda/2 \rfloor (\ell - 1)$ to $i + \lfloor \lambda/2 \rfloor \ell - 1$, and the qth subsegment contains elements of π from position $i + \lfloor \lambda/2 \rfloor (q - 1)$ to j. Note that the subsegments are defined by the elements contained in them. For example, in π the qth subsegment ends at position j but in $\pi \cdot \rho(i,j)$ it starts at position i. Even so, we can still refer to it as the qth subsegment.

The idea now is to move each subsegment until its respective position in $\pi \cdot \rho(i,j)$ by exchanging a subsegment with the one at its right. The reversals used will have size at most $2 \lfloor \lambda/2 \rfloor$, and so they are λ-reversals.

For each value of ℓ, from $\ell = 1$ up to $\ell = q - 1$, we apply a sequence of λ-reversals that first exchanges the ℓth subsegment with the $(\ell+1)$th subsegment, then exchanges the ℓth with the $(\ell + 2)$th, and so on, until it exchanges the ℓth with the qth subsegment. Note that after applying this sequence over the ℓth subsegment, it is correctly placed in its final order (relative to $\pi \cdot \rho(i,j)$) and the $(\ell+1)$th subsegment is currently starting at position i. Also, after applying the final sequence (of one λ-reversal) over the $(q-1)$th subsegment (the last one considered), subsegments $q - 1$ and q are correctly placed in their final order.

Note that exactly $q - \ell$ λ-reversals are performed over the ℓth subsegment, so a total of $(q - 1) + (q - 2) + \cdots + 1 = \frac{q(q-1)}{2}$ λ-reversals are required to correctly position all segments. Now, if q is even, then at the end of the process we will directly have $\pi \cdot \rho(i,j)$. Otherwise, all subsegments still have to be reversed and, therefore, another q reversals of size $\lfloor \lambda/2 \rfloor$ (except, maybe, for the one over the qth subsegment) are applied, one for each subsegment. Thus, the effect of a reversal can be obtained by at most $\frac{q(q-1)}{2} + q = \frac{q(q+1)}{2}$ λ-reversals.   □

As an example of the previous lemma, let $\pi = (1\,2\,3\,4\,5\,6\,7\,8)$ and suppose we want to obtain $\pi \cdot \rho(2,6)$ with 4-reversals. Let $A = (2\,3)$, $B = (4\,5)$, and $C = (6)$ be the subsegments to be moved, as described in Lemma 2. Note that the sum of the size of any two consecutive segments is less than or equal to $\lambda = 4$. The process described in the lemma first exchanges A with B, generating $(1\,5\,4\,3\,2\,6\,7\,8)$, and then with C, generating $(1\,5\,4\,6\,2\,3\,7\,8)$. Then it exchanges B with C, generating $(1\,6\,4\,5\,2\,3\,7\,8)$, and the process ends. Note that $q = \lceil (j - i + 1)/\lfloor \lambda/2 \rfloor \rceil = \lceil (6 - 2 + 1)/\lfloor 4/2 \rfloor \rceil = \lceil 5/2 \rceil = 3$ is an odd number so, although the segments are in their correct positions, we still have to revert A, B, and C (actually, note that it is not necessary to revert C, since $|C| = 1$). Thus, we obtain $\pi \cdot \rho(2,6)$ with $5 = 3 + 2 \leq q(q+1)/2 = (3 \times 4)/2 = 6$ 4-reversals.

Lemma 3. *For a permutation π and $\lambda \geq 2$, the effect of a transposition $\tau(i,j,k)$ of size $k - i > \lambda$ can be obtained by at most $\left\lceil \frac{j-i}{\lceil \lambda/2 \rceil} \right\rceil \left\lceil \frac{k-j}{\lfloor \lambda/2 \rfloor} \right\rceil$ λ-transpositions.*

Proof. Let F denote the first segment of the transposition, which contains elements of π comprised between positions i and $j - 1$, and let S denote the second segment, which contains elements of π comprised between positions j and $k - 1$. We divide F into $f = \lceil (j - i)/\lceil \lambda/2 \rceil \rceil$ subsegments of size $\lceil \lambda/2 \rceil$, except maybe for the one that ends at $j - 1$, and we divide S into $s = \lceil (k - j)/\lfloor \lambda/2 \rfloor \rceil$ subsegments of size $\lfloor \lambda/2 \rfloor$, also except maybe for the one that ends at $k - 1$, in the following manner. For $1 \leq \ell < f$, the ℓth subsegment of F contains elements of π from position $i + \lceil \lambda/2 \rceil (\ell - 1)$ to $i + \lceil \lambda/2 \rceil \ell - 1$ and the fth subsegment of F contains elements of π from position $i + \lceil \lambda/2 \rceil (f - 1)$ to $j - 1$. For $1 \leq \ell < s$, the ℓth subsegment of S contains elements of π from position $j + \lfloor \lambda/2 \rfloor (\ell - 1)$ to $j + \lfloor \lambda/2 \rfloor \ell - 1$ and the sth subsegment of S contains elements of π from position $j + \lfloor \lambda/2 \rfloor (s - 1)$ to $k - 1$. Again, note that the segments F and S and their subsegments were defined by the elements contained in them.

The idea now is to move each subsegment of F until their respective positions in $\pi \cdot \tau(i,j,k)$ by exchanging them with subsegments of S. The transpositions used will have size at most $\lceil \lambda/2 \rceil + \lfloor \lambda/2 \rfloor$ and so they are λ-transpositions.

For each value of ℓ, starting from $\ell = f$ and going down to $\ell = 1$, we apply a sequence of λ-transpositions that first exchanges the ℓth subsegment of F with the 1st segment of S, then exchanges the ℓth of F with the 2nd of S, and so on, until it exchanges the ℓth of F with the sth subsegment of S. Note that after applying this sequence of the ℓth subsegment of F, it is correctly placed in its final position (relative to $\pi \cdot \tau(i,j,k)$) and segment S is as it is in π but beginning to the right of the $(\ell - 1)$th subsegment of F. Thus, the next iteration (for $\ell - 1$) will correctly exchange the $(\ell - 1)$th subsegment of F with all subsegments of S and place it in its final position.

Note that exactly s λ-transpositions are performed over the ℓth subsegment of F, so a total of $fs = \lceil (j - i)/\lceil \lambda/2 \rceil \rceil \lceil (k - j)/\lfloor \lambda/2 \rfloor \rceil$ λ-transpositions are required to correctly position all segments.   □

Theorems 4 to 8 use Lemmas 2 and 3 to obtain approximation algorithms for the three problems we are considering.

Theorem 4. *Sorting permutations by λ-reversals has an approximation algorithm of factor $0.6875p(p+1)$, where $p = \left\lceil \frac{n}{\lfloor \lambda/2 \rfloor} \right\rceil$.*

Proof. Let ALG_r be the 1.375-approximation algorithm for sorting permutations by reversals [2]. We can create an algorithm for sorting permutations by λ-reversals by changing each reversal given by ALG_r over a permutation π for a sequence of λ-reversals, as described in Lemma 2. Since a reversal can have size at most n, it will be replaced by at most $p = \lceil n/\lfloor \lambda/2 \rfloor \rceil$ λ-reversals.

Let $\text{ALG}_r(\pi)$ be the size of the sorting sequence produced by ALG_r to sort a permutation π. The amount of λ-reversals used by our algorithm to sort π is thus at most $\text{ALG}_r(\pi)\frac{p(p+1)}{2} \leq 1.375 d_r(\pi)\frac{p(p+1)}{2} \leq 0.6875p(p+1)d_r^\lambda(\pi)$, where the last inequality follows from Lemma 1. □

Corollary 5. *Sorting permutations by λ-reversals has an approximation algorithm of factor 8.25 for all $n > 3$ and $\lambda > \lceil n/2 \rceil$.*

Theorem 6. *Sorting permutations by λ-transpositions has an approximation algorithm of factor $1.375 \left\lceil \frac{\lceil n/2 \rceil}{\lceil \lambda/2 \rceil} \right\rceil \left\lceil \frac{\lfloor n/2 \rfloor}{\lfloor \lambda/2 \rfloor} \right\rceil$.*

Proof. Let ALG_t be the 1.375-approximation algorithm for sorting permutations by transpositions [7]. We can create an algorithm for sorting permutations by λ-transpositions by changing each transposition given by ALG_t for a sequence of λ-transpositions, as described in Lemma 3. Since a transposition can have size at most n, it will be replaced by at most $T = \lceil \lceil n/2 \rceil / \lceil \lambda/2 \rceil \rceil \lceil \lfloor n/2 \rfloor / \lfloor \lambda/2 \rfloor \rceil$ λ-transpositions.

Let $\text{ALG}_t(\pi)$ be the size of the sorting sequence produced by ALG_t to sort a permutation π. The amount of λ-transpositions used by our algorithm to sort π is thus at most $\text{ALG}_t(\pi)\,T \leq 1.375\,T\,d_t(\pi) \leq 1.375\,T\,d_t^\lambda(\pi)$, where the last inequality follows from Lemma 1. □

Corollary 7. *Sorting permutations by λ-transpositions has an approximation algorithm of factor 5.5 for all $n > 3$ and $\lambda > \lceil n/2 \rceil$.*

Theorem 8. *Sorting permutations by λ-reversals and λ-transpositions has an approximation algorithm of factor $\alpha\,p(p+1)$, where α is the approximation factor of the cycle decomposition algorithm for breakpoint graphs and $p = \left\lceil \frac{n}{\lfloor \lambda/2 \rfloor} \right\rceil$.*

Proof. Let ALG_{rt} be the 2α-approximation algorithm for sorting permutations by reversals and transpositions [15]. Our algorithm for sorting permutations by λ-reversals and λ-transpositions transforms the sequence of operations given by ALG_{rt} into λ-operations, as described in Lemmas 2 and 3.

Since reversals and transpositions can have size at most n, each transposition is replaced by at most $T = \lceil \lceil n/2 \rceil / \lceil \lambda/2 \rceil \rceil \lceil \lfloor n/2 \rfloor / \lfloor \lambda/2 \rfloor \rceil$ λ-transpositions, and each reversal is replaced by at most $R = 1/2(\lceil n/\lfloor \lambda/2 \rfloor \rceil (\lceil n/\lfloor \lambda/2 \rfloor \rceil + 1))$ λ-reversals. As $T \leq R$, we can suppose that, in the worst case, ALG_{rt} only uses reversals to sort. Let $\text{ALG}_{rt}(\pi)$ be the size of the sorting sequence used by ALG_{rt}

to sort a permutation π. The amount of operations used by our algorithm is at most $\mathrm{ALG}_{rt}(\pi)R \leq 2\alpha R\, d_{rt}(\pi) \leq \alpha\, p(p+1)d_{rt}^{\lambda}(\pi)$, where the last inequality follows from Lemma 1. □

Corollary 9. *Sorting permutations by λ-reversals and λ-transpositions has an approximation algorithm of factor 12α for all $n > 3$ and $\lambda > \lceil n/2 \rceil$.*

3.2   Approximation Algorithms for Small Values of λ

In this section we present algorithms that have a better approximation factor when the value of λ is small. To do this, we first need some new definitions.

The *entropy of an element* π_i, denoted by $\mathrm{ent}(\pi_i)$, is given by $|\pi_i - i|$, that is, the distance between π_i and its position in ι. The *entropy of a permutation* π, denoted by $\mathrm{ent}(\pi)$, is given by the sum of all values of $\mathrm{ent}(\pi_i)$, for $1 \leq i \leq n$. For example, the entropy of $\pi = (2\ 3\ 5\ 4\ 1)$ is $\mathrm{ent}(\pi) = 1 + 1 + 2 + 0 + 4 = 8$. Note that, by definition, the entropy of any permutation is an even number. Also note that $\mathrm{ent}(\iota) = 0$ and $\mathrm{ent}(\pi) > 0$ for all $\pi \neq \iota$.

Let $\Delta_{\mathrm{ent}}(\pi, \sigma) = \mathrm{ent}(\pi \cdot \sigma) - \mathrm{ent}(\pi)$ be the variation of the entropy after the application of an operation σ. Note that, to calculate $\Delta_{\mathrm{ent}}(\pi, \sigma)$, it suffices to determinate the entropy of the elements affected by σ (the entropy of other elements does not change).

Lemma 10 gives an upper bound on the variation of entropy caused by any λ-reversal or λ-transposition.

Lemma 10. *Let $\Delta_{\mathrm{ent}}^{max}(\pi, \sigma)$ be the maximum variation of entropy caused by a λ-reversal or by a λ-transposition σ. We have $\Delta_{\mathrm{ent}}^{max}(\pi, \sigma) = 2 \lceil \lambda/2 \rceil \lfloor \lambda/2 \rfloor$.*

Proof. First let σ be a λ-reversal $\rho(i, j)$. Note that the maximum variation of entropy occurs when $j - i + 1 = \lambda$, when the most amount of elements are involved. In this case, after the reversal, element π_i ends up $\lambda - 1$ positions away from i, element π_{i+1} ends up $\lambda - 3$ positions away from $i + 1$, and so on, up to element $\pi_{i+\lfloor \lambda/2 \rfloor - 1}$, which ends up 1 position away from $i + \lfloor \lambda/2 \rfloor - 1$, if λ is even, or it ends up 2 positions away from $i + \lfloor \lambda/2 \rfloor - 1$, if λ is odd. Similarly, element π_j ends up $\lambda - 1$ positions away from j, element π_{j-1} ends up $\lambda - 3$ positions away from $j - 1$, and so on, down to element $\pi_{j-\lfloor \lambda/2 \rfloor + 1}$, which similarly ends up 1 or 2 positions away from $j - \lfloor \lambda/2 \rfloor + 1$, if λ is even or odd, respectively. When λ is odd, element $i + \lceil \lambda/2 \rceil$ remains in the same position. Therefore, the maximum variation of entropy for each element between positions i and j is at most $2\sum_{\ell=1}^{\lfloor \lambda/2 \rfloor}(\lambda - (2\ell - 1)) = 2 \lfloor \lambda/2 \rfloor \lceil \lambda/2 \rceil$.

Now let σ be a λ-transposition $\tau(i, j, k)$ and, similarly, suppose $k - i = \lambda$ so we can have the maximum variation of entropy. In this case, after the transposition, all elements π_h, for $i \leq h < j$, end up $k - j$ positions away from h, while all elements π_ℓ, for $j \leq \ell < k$, end up $j - i$ positions away from ℓ. Since there are $j - i$ elements of the first type and $k - j$ elements of the second type, the maximum variation of entropy for each element between i and k is at most $2(j - i)(k - j) \leq 2 \lfloor \lambda/2 \rfloor \lceil \lambda/2 \rceil$. □

Corollary 11. *For any permutation π, $\lambda \geq 2$, and $\beta \in \{r, t, rt\}$, we have* $d_\beta^\lambda(\pi) \geq \text{ent}(\pi)/(2\lceil\lambda/2\rceil\lfloor\lambda/2\rfloor)$.

Let π be a permutation of size n and let i and j be two integers such that $1 \leq i < j \leq n$. Function $\phi(\pi, i, j)$ returns a permutation π' such that $\pi'_i = \pi_j$, $\pi'_j = \pi_i$, and $\pi'_k = \pi_k$ for all $k \notin \{i, j\}$. In other words, only elements π_i and π_j are exchanged. Lemmas 12 and 13 show how to obtain $\phi(\pi, i, j)$ with λ-reversals and λ-transpositions, respectively.

Lemma 12. *Let π be a permutation, $\lambda \geq 2$, and i and j be positions such that $1 \leq i < j \leq n$. It is possible to obtain $\phi(\pi, i, j)$ by applying at most $2x$ λ-reversals on π, where $x = \left\lceil \frac{j-i}{\lambda-1} \right\rceil$.*

Proof. We show that the result follows by considering two cases, according to the relation between i and j. If $j - i \leq \lambda - 1$, then at most two λ-reversals are necessary. First we apply the operation $\rho(i, j)$ that exchanges the position of elements π_i and π_j. It is easy to see that, if $j - i + 1 \leq 3$, we already obtain $\phi(\pi, i, j)$ with only this operation. Otherwise, observe that we will have the segment $\pi_{i+1}, \ldots, \pi_{j-1}$ in reverse order (regarding $\phi(\pi, i, j)$). Thus, we have to apply a second operation $\rho(i + 1, j - 1)$ to revert it again and, then, we obtain $\phi(\pi, i, j)$ with two λ-reversals.

Otherwise, $j - i > \lambda - 1$. In a first step, we move element π_i to position j by repeatedly increasing its position by $\lambda - 1$ (except, maybe, at the last movement) with exactly $x = \lceil (j - i)/(\lambda - 1) \rceil$ λ-reversals applied successively. Formally, this is done by applying the sequence $\rho(i, i + (\lambda - 1))$, $\rho(i + (\lambda - 1), i + 2(\lambda - 1))$, $\rho(i + 2(\lambda - 1), i + 3(\lambda - 1))$, $\ldots$, $\rho(i + (x - 1)(\lambda - 1), j)$ of λ-reversals. Now element π_j is at position $i + (x - 1)(\lambda - 1)$ and elements π_t, for $i < t < j$, are not necessarily at position t. To correct this and, at the same time, move element π_j to position i, we have a second step that applies a sequence with the same λ-reversals that were used before (except for the last λ-reversal) in reversed order, to repeatedly decrease the position of π_j by $\lambda - 1$. Thus, a total of $x - 1$ extra operations are needed. Formally, the sequence is $\rho(i + (x - 2)(\lambda - 1), i + (x - 1)(\lambda - 1))$, $\rho(i + (x - 3)(\lambda - 1), i + (x - 2)(\lambda - 1))$, $\ldots$, $\rho(i + 2(\lambda - 1), i + 3(\lambda - 1))$, $\rho(i + (\lambda - 1), i + 2(\lambda - 1))$, $\rho(i, i + (\lambda - 1))$. At this point, if the size of the λ-reversal $\rho(i + (x - 1)(\lambda - 1), j)$ (the last λ-reversal of the first step) is less than or equal to 3, then at the end of the process we will directly have $\phi(\pi, i, j)$. Otherwise, we have to apply one more λ-reversal $\rho(i + (x - 1)(\lambda - 1) + 1, j - 1)$ to obtain $\phi(\pi, i, j)$, which totalizes $2x$ λ-reversals.   □

Lemma 13. *Let π be a permutation, $\lambda \geq 2$, and i and j be positions such that $1 \leq i < j \leq n$. It is possible to obtain $\phi(\pi, i, j)$ by applying exactly $x + y$ λ-transpositions on π, where $x = \left\lceil \frac{j-i}{\lambda-1} \right\rceil$ and $y = \left\lceil \frac{j-i-1}{\lambda-1} \right\rceil$.*

Proof. We show that the result follows by considering two cases, according to the relation between i and j. If $j - i \leq \lambda - 1$, then at most two λ-transpositions are necessary. First we apply $\tau(i, i + 1, j + 1)$ that puts element π_i at position j.

When $j = i+1$, it also puts element π_j at position i and, since there is no elements between π_i and π_j, we have already obtained $\phi(\pi, i, j)$. Otherwise, observe that we will have elements π_t, for $i < t \leq j$, exactly one position to the left of their original position in π. Thus, we have to apply a second operation $\tau(i, j-1, j)$ in order to correct this. Note that, after applying this second operation, we have π_j at position i at the same time that elements $i < t < j$ were moved one position to the right and, then, we got $\phi(\pi, i, j)$ with two λ-transpositions.

Otherwise, $j - i > \lambda - 1$. Initially we move element π_i to position j by repeatedly increasing its position by $\lambda - 1$ (except, maybe, at the last movement) with exactly $x = \lceil (j - i)/(\lambda - 1) \rceil$ λ-transpositions applied successively. Formally, this is done by applying the sequence of λ-transpositions $\tau(i, i+1, i+\lambda)$, $\tau(i+(\lambda-1), i+(\lambda-1)+1, i+2(\lambda-1))$, $\tau(i+2(\lambda-1), i+2(\lambda-1)+1, i+2(\lambda-1)+\lambda)$, ..., $\tau(i+(x-1)(\lambda-1), i+(x-1)(\lambda-1)+1, j+1)$. After this, each element π_t, for $i < t \leq j$, is exactly one position to the left of its original position in π. To correct this, we can apply a similar sequence of λ-transpositions, but now we will repeatedly decrease the position of π_j (which is at position $j - 1$) by $\lambda - 1$ (except, maybe, at the last operation) with exactly $y = \lceil (j - 1 - i)/(\lambda - 1) \rceil$ λ-transpositions applied successively. Formally, the sequence is $\tau(j - \lambda, j - 1, j)$, $\tau(j - (\lambda - 1) - \lambda, j - (\lambda - 1) - 1, j - (\lambda - 1))$, $\tau(j - 2(\lambda - 1) - \lambda, j - 2(\lambda - 1) - 1, j - 2(\lambda - 1))$, ..., $\tau(i, j - y(\lambda - 1) - 1, j - y(\lambda - 1))$. Note that each λ-transposition moves $\lambda - 1$ (again, except, maybe, the last operation) elements π_t, for $i < t < j$, one position to the right by exchanging all of them with element π_j. At the end of this process we directly have $\phi(\pi, i, j)$. $\square$

Since $\text{ent}(\iota) = 0$ and $\text{ent}(\pi) > 0$ for all $\pi \neq \iota$, an algorithm that always reduces the entropy of the input permutation will eventually reach the identity. Lemma 14 is auxiliar to Lemma 15, which shows that it is always possible to reduce the entropy of any permutation.

Lemma 14. *For all permutations $\pi \neq \iota$, there exists a pair of elements π_i and π_j, with $1 \leq i < j \leq n$, such that $\pi_i \geq j$ and $\pi_j \leq i$.*

Proof. Let G_π be the directed graph such that $V(G_\pi) = \{1, 2, \ldots, n\}$ and $E(G_\pi) = \{(\pi_i, i) : 1 \leq i \leq n\}$. Note that each vertex has in-degree 1 and out-degree 1, and, therefore, the components of G_π are cycles. Also note that only G_ι has n unitary cycles.

Let C be any cycle of G_π with at least two vertices and let u be the smallest-value vertex of C. Let $B = (v_1, v_2, \ldots, v_\ell)$ be a maximal sequence of vertices of C such that $v_1 = u$, $v_i < v_{i+1}$ for all $1 \leq i < \ell$, and $(v_i, v_{i+1}) \in E(G_\pi)$.

Since the vertices of B are in a cycle and B is maximal, the edge incident to v_ℓ is of the form (v_ℓ, x), with $x < v_\ell$. If $v_{\ell-1} \leq x$, then take $i = x$ and $j = v_\ell$. In this case, we have $\pi_i = \pi_x = v_\ell = j$ and $\pi_j = \pi_{v_\ell} = v_{\ell-1} \leq x = i$ and the lemma follows. If $v_{\ell-1} > x$, then let k, for $1 \leq k < \ell - 1$, be such that $v_k \leq x < v_{k+1}$ and take $i = x$ and $j = v_{k+1}$. In this case, we have $\pi_i = \pi_x = v_\ell > v_{k+1} = j$ and $\pi_j = \pi_{v_{k+1}} = v_k \leq x = i$ and the lemma follows. See Fig. 1 for an example. $\square$

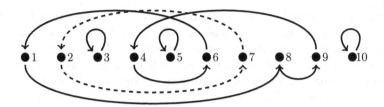

Fig. 1. Graph G_π for $\pi = (6\,7\,3\,9\,5\,4\,2\,1\,8\,10)$. Considering the notation of Lemma 14's proof, if $C = (1, 8, 9, 4, 6, 1)$, then $B = (1, 8, 9)$.

Lemma 15. *For all permutations π, it is possible to obtain another permutation $\phi(\pi, i, j)$ such that $\mathrm{ent}(\phi(\pi, i, j)) = \mathrm{ent}(\pi) - 2(j - i)$, for some pair $1 \le i < j \le n$.*

Proof. Let i and j be as in Lemma 14. By definition, $\mathrm{ent}(\pi) - \mathrm{ent}(\phi(\pi, i, j)) = |\pi_i - i| + |\pi_j - j| - |\pi_i - j| - |\pi_j - i|$. Since $\pi_i \ge j > i$, we have $|\pi_i - j| = \pi_i - j$ and $|\pi_i - i| = \pi_i - i$. Since $\pi_j \le i < j$, we have $|\pi_j - i| = i - \pi_j$ and $|\pi_j - j| = j - \pi_j$. Thus, $\mathrm{ent}(\pi) - \mathrm{ent}(\phi(\pi, i, j)) = \pi_i - i + j - \pi_j - \pi_i + j - i + \pi_j = 2(j - i)$. □

A permutation $\pi \ne \iota$ is called *normal* if there exists one λ-operation σ such that $\mathrm{ent}(\pi \cdot \sigma) < \mathrm{ent}(\pi)$. Moreover, since the entropy of any permutation is an even number, $\mathrm{ent}(\pi) - \mathrm{ent}(\pi \cdot \sigma) \ge 2$. Otherwise π is called *special*.

Consider the unsigned normal permutation $\pi = (4\,5\,2\,3\,1)$ whose entropy is 12. Note that $\mathrm{ent}(\pi \cdot \rho(2, 3)) = 10$ but $\mathrm{ent}(\pi \cdot \rho(1, 5)) = 4$. The following algorithm is greedy in the sense of always choosing a λ-operation that most decreases the entropy when applied to a normal permutation. When the permutation is special, Lemma 15 is applied. The algorithm receives the rearrangement model β, since it works for any of the problems we are considering, and it runs in polynomial time, as we discuss later. We show its approximation factor in Theorem 16.

```
function GreedyAlgorithm(π, λ, β):
    while ent(π) > 0 do
        if π is a normal permutation then
            Let σ be a λ-operation such that ent(π) − ent(π · σ) is maximum;
            π ← π · σ;
        else
            Let i and j be positions such that πᵢ ≥ j, πⱼ ≤ i, and 1 ≤ i < j ≤ n;
            π ← φ(π, i, j), according to Lemma 12 or 13;
        end
    end
```

Theorem 16. *Sorting permutations by λ-reversals, by λ-transpositions, or by both operations has an approximation algorithm of factor $4 \lceil \lambda/2 \rceil \lfloor \lambda/2 \rfloor$.*

Proof. If π is a normal permutation, then the amount of entropy decreased per operation is at least 2, as mentioned above.

Otherwise π is special. Let i and j be as in GREEDYALGORITHM and note that $\pi' = \phi(\pi, i, j)$ is a permutation with $\text{ent}(\pi') = \text{ent}(\pi) - 2(j - i)$ (see Lemma 15). Also, we obtain π' through the operations described in Lemma 12 or 13.

If $j - i \leq \lambda - 1$, then we obtain π' by applying at most two λ-operations. Therefore, the amount of entropy decreased per operation is at least $(2(j - i))/2 \geq 1$, for any of the 3 problems.

If $j - i > \lambda - 1$, then we obtain π' by applying at most $2 \lceil (j - i)/(\lambda - 1) \rceil$ λ-reversals or $\lceil (j - i)/(\lambda - 1) \rceil + \lceil (j - i - 1)/(\lambda - 1) \rceil$ λ-transpositions. Since $2 \lceil (j - i)/(\lambda - 1) \rceil \geq \lceil (j - i)/(\lambda - 1) \rceil + \lceil (j - i - 1)/(\lambda - 1) \rceil$, we have the amount of entropy decreased per operation is at least $\frac{2(j-i)}{2\lceil \frac{j-i}{\lambda-1} \rceil} = \frac{j-i}{\lceil \frac{j-i}{\lambda-1} \rceil} \geq \frac{j-i}{2\frac{j-i}{\lambda-1}} = \frac{\lambda-1}{2} \geq \frac{1}{2}$.

Thus, in the worst case we only reduce the entropy by $1/2$ per operation for all the problems we are addressing, which means that the total amount of operations used by GREEDYALGORITHM is at most $\frac{\text{ent}(\pi)}{1/2} = 2\,\text{ent}(\pi) \leq 2\,d_r^\lambda(\pi)\,2\,\lceil \lambda/2 \rceil \lfloor \lambda/2 \rfloor$, where the inequality follows from Lemma 10. $\qquad\square$

Regarding the time complexity of the algorithm, first observe that the while loop at the first line runs at most $O(n^2)$ times because the entropy of a permutation is $O(n^2)$ and, as seen in Theorem 16, one operation reduces the entropy by at least $1/2$. In the "if" command we must decide whether a permutation is normal, which can be done by finding the desired λ-operation. We can simply test all possible λ-operations, which takes time $O(n\lambda)$ for λ-reversals and $O(n\lambda^2)$ for λ-transpositions. Furthermore, it takes $O(\lambda)$ time to calculate the entropy's variation. Therefore, testing if a permutation is normal takes time $O(n\lambda^3)$. In the "else" command, we must first obtain positions i and j as desired, which takes time $O(n)$. Now consider obtaining $\phi(\pi, i, j)$ with λ-transpositions. By Lemma 13, we can use at most $\lceil (j - i)/(\lambda - 1) \rceil + \lceil (j - i - 1)/(\lambda - 1) \rceil$ such operations, each one of size at most λ, which means the time to perform each transposition is $O(\lambda)$ and so the total time to obtain $\phi(\pi, i, j)$ is at most $O(\lambda)\left(\frac{j-i+1}{\lambda-1} + \frac{j-i}{\lambda-1}\right) \leq O(\lambda)2\frac{n}{\lambda-1} = O(n\lambda)$, since $\lambda \geq 2$. Similarly, the time to obtain $\phi(\pi, i, j)$ with reversals is also $O(n\lambda)$. Therefore, the total time of GREEDYALGORITHM is $O(n^3\lambda^3)$ for any of the problems we are considering. Note that this is polynomial because $\lambda = O(n)$.

4   Conclusion

We have presented six approximation algorithms for the problems of sorting permutations by λ-reversals and/or λ-transpositions, being 2 algorithms for each problem we considered, where one of them works better when we have large values of λ and the other one works better when we have small values of λ. We are currently extending these results to signed permutations, in which each element is associated with a sign '+' or '−'. These permutations are used to represent the genomes when the orientation of the genes is known, and so the sign of each element is used to indicate this.

Acknowledgments. This work was supported by the Brazilian Federal Agency for the Support and Evaluation of Graduate Education, CAPES, the National Counsel of Technological and Scientific Development, CNPq (grants 400487/2016-0, 425340/2016-3, and 131182/2017-0), the São Paulo Research Foundation, FAPESP (grants 2013/08293-7, 2015/11937-9, 2016/14132-4, and 2017/12646-3), and the program between CAPES and the French Committee for the Evaluation of Academic and Scientific Cooperation with Brazil, COFECUB (grant 831/15).

References

1. Bafna, V., Pevzner, P.A.: Sorting by transpositions. SIAM J. Discret. Math. **11**(2), 224–240 (1998)
2. Berman, P., Hannenhalli, S., Karpinski, M.: 1.375-approximation algorithm for sorting by reversals. In: Möhring, R., Raman, R. (eds.) ESA 2002. LNCS, vol. 2461, pp. 200–210. Springer, Heidelberg (2002). https://doi.org/10.1007/3-540-45749-6_21
3. Bulteau, L., Fertin, G., Rusu, I.: Sorting by transpositions is difficult. SIAM J. Comput. **26**(3), 1148–1180 (2012)
4. Caprara, A.: Sorting permutations by reversals and Eulerian cycle decompositions. SIAM J. Discret. Math. **12**(1), 91–110 (1999)
5. Chen, T., Skiena, S.S.: Sorting with fixed-length reversals. Discret. Appl. Math. **71**(1–3), 269–295 (1996)
6. Dias, Z., Meidanis, J.: Sorting by prefix transpositions. In: Laender, A.H.F., Oliveira, A.L. (eds.) SPIRE 2002. LNCS, vol. 2476, pp. 65–76. Springer, Heidelberg (2002). https://doi.org/10.1007/3-540-45735-6_7
7. Elias, I., Hartman, T.: A 1.375-approximation algorithm for sorting by transpositions. IEEE/ACM Trans. Comput. Biol. Bioinf. **3**(4), 369–379 (2006)
8. Galvão, G.R., Lee, O., Dias, Z.: Sorting signed permutations by short operations. Algorithms Mol. Biol. **10**(1), 1–17 (2015)
9. Heath, L.S., Vergara, J.P.C.: Sorting by short swaps. J. Comput. Biol. **10**(5), 775–789 (2003)
10. Jerrum, M.R.: The complexity of finding minimum-length generator sequences. Theoret. Comput. Sci. **36**(2–3), 265–289 (1985)
11. Jiang, H., Feng, H., Zhu, D.: An 5/4-approximation algorithm for sorting permutations by short block moves. In: Ahn, H.-K., Shin, C.-S. (eds.) ISAAC 2014. LNCS, vol. 8889, pp. 491–503. Springer, Cham (2014). https://doi.org/10.1007/978-3-319-13075-0_39
12. Lefebvre, J.F., El-Mabrouk, N., Tillier, E.R.M., Sankoff, D.: Detection and validation of single gene inversions. Bioinformatics **19**(1), i190–i196 (2003)
13. Lin, G., Jiang, T.: A further improved approximation algorithm for breakpoint graph decomposition. J. Comb. Optim. **8**(2), 183–194 (2004)
14. Lintzmayer, C.N., Fertin, G., Dias, Z.: Sorting permutations by prefix and suffix rearrangements. J. Bioinf. Comput. Biol. **15**(1), 1750002 (2017)
15. Rahman, A., Shatabda, S., Hasan, M.: An approximation algorithm for sorting by reversals and transpositions. J. Discret. Algorithms **6**(3), 449–457 (2008)
16. Vergara, J.P.C.: Sorting by bounded permutations. Ph.D. thesis, Virginia Polytechnic Institute and State University (1998)
17. Walter, M.E.M.T., Dias, Z., Meidanis, J.: Reversal and transposition distance of linear chromosomes. In: Proceedings of the 5th International Symposium on String Processing and Information Retrieval (SPIRE 1998), pp. 96–102. IEEE Computer Society, Los Alamitos (1998)

Fast Algorithm for Vernier Search of Long Repeats in DNA Sequences with Bounded Error Density

Sergey P. Tsarev[1], Maria Y. Senashova[2], and Michael G. Sadovsky[1,2](✉)

[1] Siberian Federal University,
Svobodny prosp., 79, 660049 Krasnoyarsk, Russia
sptsarev@mail.ru
[2] Institute of computational modelling of SB RAS,
Akademgorodok, 660036 Krasnoyarsk, Russia
{msen,msad}@icm.krasn.ru,
http://icm.krasn.ru

Abstract. A novel algorithm to find all sufficiently long repeating nucleotide substrings in one or several DNA sequences is proposed. The algorithm searches approximately matching strings very fast with given level of local error density. Some biological applications illustrate the method.

Keywords: Multiple sequence alignment · Edit distance · Gauge
Repeat · Insertion · Deletion · Mutation

1 Introduction

The search of common strings in two or several symbol sequences makes a core in bioinformatics and up-to-date molecular biology [1,7]. The problem is far from a completion, in spite of a long story [2,9,12]. In general, the problem is the following: given sequences $\mathfrak{T}_1, \mathfrak{T}_2, \ldots, \mathfrak{T}_k$ of symbols from some finite alphabet, find all possible common substrings (i.e. coherent subsequences) occurred in the sequences $\mathfrak{T}_i$, maybe, with some mismatches. Further we shall concentrate on the four-letter alphabet $\aleph = \{A, C, G, T\}$.

Previously, a new algorithm for the fast search of common substrings in two or several symbol sequences had been reported [15]. The algorithm was originally implemented for the exact matching strings search, while it allows some extensions for error tolerant search of substrings. The algorithm [15] for search of exactly matching substrings is much faster compared to the brute force search methods; it is based on a simple idea of rarefied dictionaries and uses the classical Vernier scale, cf. for example [16].

Here we provide the modification of Vernier gauge algorithm [15] that is extended for the case of errors of *deletion* and *insertion* types. Theoretical background of the algorithm correctness is provided, as well as some computational

J. Jansson et al. (Eds.): AlCoB 2018, LNBI 10849, pp. 88–99, 2018.
https://doi.org/10.1007/978-3-319-91938-6_8

results showing its high speed and efficiency, for the case of genetic sequences analysis. The modification allows all three types of errors (these are mutation, insertion and deletion). It should be said that the most popular BLAST engine [2,3], and other up-to-date methods (see, e.g. [5,10]) may also resolve the problem. The point is that any homology search consists (roughly) of two steps: the former is *seeds* search of exactly matching k-mers (executed over the complete dictionary), and the latter is the expansion of them. The method proposed here does not, yet, compete to BLAST in expansion procedure; our method takes an advantage over BLAST in the speed of the *seeds* search. The advantage results from the rarefication of dictionary of k-mers, thus accelerating the search procedure. Also, an additional advantage of our *seeds* search algorithm consists in significantly reduced number of false (not expandable) matching *seeds*, and giving a possibility to increase the length k of k-mers. Another important advantageous issue of the proposed algorithm is that it allows mismatches and indels in *seed* search, unlike classic BLAST. We found 30 times acceleration, in our experiments, at the search step.

Let us introduce some notions and definitions; $\mathfrak{T}_1$, $\mathfrak{T}_2$, ..., $\mathfrak{T}_k$ make the set of sequences from four-letter alphabet $\aleph = \{\mathsf{A}, \mathsf{C}, \mathsf{G}, \mathsf{T}\}$. S_j, $1 \leq j \leq k$ are the "common" parts (substrings), so that $S_j \subset \mathfrak{T}_j$, and $l_j = l(S_j) = |S_j|$, $1 \leq j \leq k$ are the lengths of those "common" substrings. The substrings $S_j \subset \mathfrak{T}_j$ may not be identical, and if it happens, then they must be *approximately matching*; the exact definition of this approximate matching see below in Sect. 3 (we adopt the classical definition of *edit distance* to measure substring mismatching [8,11]). Similarly, they may vary in length, while the variation must not be too large. Frequency dictionary $W_{(q,t)}$ is the set of the substrings of the fixed length q within a sequence $\mathfrak{T}$ so, that the window of the given length q moves along the sequence $\mathfrak{T}$ with the step t; see details in [4, 13–15].

Section 2 briefly presents the previous version of our algorithm targeted to search the exactly matching substrings. Section 3 provides the description of the new algorithm feasible for any type of mismatches in the strings under comparison as well, as necessary theoretical issues with rigorous estimations for algorithm parameters. Finally, Sect. 4 shows some computational experiments proving the efficiency and feasibility of the method and its implementation. Everywhere below, the classical BLAST *seeds* are called *tags*.

2   Fast Vernier Search Algorithm for Exact Matching

For better explanation of our new algorithm of approximate matching search, we briefly consider the algorithm of the exact matching search presented in [15]. This is how the method works. Suppose, there are two sequences $\mathfrak{T}_1$ and $\mathfrak{T}_2$ each having the common substring S (here S is the same in two sequences), and N is the lower bound of the expected length of S.

Step 1. Given the target length bound N, choose as great k, as possible and m (in practice m varies from 20 to 50, see details in [15]) such that $N \geq k(k-1) + m - 1$.

Step 2. Develop two dictionaries $W^{[1]}_{m,k}$ and $W^{[2]}_{m,k-1}$, over sequences $\mathfrak{T}_1$ and $\mathfrak{T}_2$, respectively. Any substring of the length m in these dictionaries is a tag, we also keep the positions of the tags in the dictionaries. This construction implements Vernier gauge algorithm described in detail in [15].

Step 3. Find identical tags in these two dictionaries $W^1_{m,k}$ and $W^2_{m,k-1}$ (for example using lexicographic sorting) and return the common tags (including positions); if the common substring S exists, so that $S \subset \mathfrak{T}_1$ and $S \subset \mathfrak{T}_2$, and k is chosen in proper way (as described above), then there must be such identical tag $W^1_{m,k}$ and $W^2_{m,k-1}$ (see Theorem 1 in [15]).

Step 4. Starting from the positions of these common tags in $\mathfrak{T}_i$, expand the tag substrings left and right letter by letter, checking every expansion step whether two expansions still coincide.

The details of that procedure see in [15]. It should be stressed that the algorithm presented in [15] supports the error tolerant search, while mutations only are permitted. Below we generalize it for the case of insertion/deletion errors.

3   New Vernier Gauge Algorithm for Insertion/Deletion Errors

Here we introduce a new method to search *approximately matching repeating substrings of the length* $\geq N$ *in a symbol sequence* $\mathfrak{T}$ *(alternatively, approximately matching substrings in several symbol sequences* $\mathfrak{T}_i$*)* generalizing the Vernier gauge approach described in Sect. 2. The key idea (similar to that in [15]) is to change the analysis of a complete dictionary $W_{m,1}$ (where each symbol in a sequence $\mathfrak{T}$ gives a start to a *tag*, i. e. substring of the length m) to the analysis of appropriately rarefied dictionary W_{m,t_s} with *variable step* t_s yielding significantly less number of entries. First let us fix the necessary definitions.

Definition 1. *Given two strings* S_1, S_2 *we call their* edit transformation *a fixed series of one-symbol mutations (substitutes of some symbol by another from the same alphabet), symbol deletions and symbol insertions transforming* S_1 *into* S_2 *and having minimal possible total number of mutations, deletions and insertions (this total number is the standard* edit distance *[8, 11]).*

As soon, as the edit transformation from S_1 into S_2 is fixed, we can keep track of the positions in S_1 of the symbols unchanged by the edit transformation; it results in a partial mapping from S_1 to S_2 that yields a correspondence of such unchanged symbols in S_1 and S_2. Let us call this correspondence a *matching map* $M_{S_1,S_2} \colon S_1 \rightleftarrows S_2$. This is a partial mapping from S_1 to S_2 and from S_2 to S_1. An example of such matching map is shown in Fig. 1.

Sometimes the edit transformation mapping S_1 to S_2 may have very long exactly matching substrings and all deletions/insertions/mutations may be concentrated in some relatively short parts of the substrings S_1, S_2. Thus, one can not distinguish between this case and the case where one has in fact *two pairs* of (approximately) matching substrings $S_{11} \subset S_1$, $S_{12} \subset S_2$ and $S_{21} \subset S_1$,

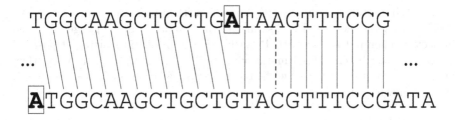

Fig. 1. A matching map. Bold letters in boxes are the deletions/insertions; the dashed line shows mutation

$S_{22} \subset S_2$ with zero (or very small) edit distance that are occasionally close to each other. To avoid such questionable situations we constrain the edit transformations: suppose that the *local density* of edits is bounded from below by a prescribed non-negative number $\delta < 1$ ($\delta = 0$ means exact matching):

Definition 2. *Given two approximately matching strings S_1, S_2, the corresponding matching map $M_{S_1,S_2}: S_1 \rightleftarrows S_2$ and an integer m, we call the* local error m-density *of this matching map the minimal non-negative number $\delta < 1$ such that any substring $s \subset S_1$ of length at least m and any M_{S_1,S_2}-corresponding substring $\widehat{s} \subset S_2$ has at most $d = \delta \cdot \text{length}(s)$ mismatches (mutations, deletions, insertions).*

Note that if the substring $s \subset S_1$ starts with a symbol (or several symbols) deleted/inserted in the corresponding edit transformation from S_1 into S_2, the M_{S_1,S_2}-corresponding substring $\widehat{s} \subset S_2$ may be determined in a number of ways; we check *all* such possible $\widehat{s}$ for the local error m-density bound δ. In fact this means the following obvious fact: if deletions/insertions are concentrated around some position, the rest of the m-neighborhood of this location must have much lower m-density of errors.

3.1 General Description of the Problem of Vernier Search of Long Repeats with Bounded Local Error Density. Modified Vernier Gauge and the Idea of the Algorithm

The general problem solved by our new algorithms is:

Given parameters N (an integer) and δ (a non-negative real number, $\delta < 0.5$), find all couples of substrings S_k of the length at least N in one or several symbol sequences $\mathfrak{T}_i$ that differ at most by the edit distance $d = \delta \cdot \max(\text{length}(S_k))$, with local error v-density $\leq \delta$, for some convenient v.

The idea to search approximately matching substrings (including *deletions and insertions*) using relatively short tags and appropriately rarefied dictionaries (we call it *Vernier gauges* on $\mathfrak{T}_i$) is based on the following simple construction. Suppose we have two symbol sequences $\mathfrak{T}_1$, $\mathfrak{T}_2$ with two approximately matching substrings $S_1 \subset \mathfrak{T}_1$, $S_2 \subset \mathfrak{T}_2$ of lengths $\geq N$ with local error m-density $\leq \delta$.

Put (virtual) marks on over the string $\mathfrak{T}_1$ with step k (we choose $k = \lfloor\sqrt{N}\rfloor$ for simplicity) as shown in Fig. 2 similar to the standard Vernier gauge described in Sect. 2. Then put $k+1$ (virtual) marks with step 1 in $\mathfrak{T}_2$ starting at position 1, then another batch of $k+1$ marks with step 1 starting at position $N-k$, then another batch of $k+1$ marks with step 1 starting at position $2(N-k-1)+1$, etc., as shown on Fig. 2.

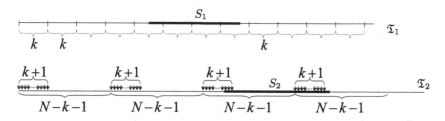

Fig. 2. Positions of tags for our algorithm of the approximate substring matching

Now we can form the rarefied dictionaries: $\widetilde{W}_m^{[1]}$ is built choosing substrings of length m (such substring are called *tags*) starting at the marked positions in the first string $\mathfrak{T}_1$ (this is in fact $W_{m,k}^{[1]}$ with constant step k) and the second dictionary $\widetilde{W}_m^{[2]}$ with variable step, choosing substrings of length m starting at the marked positions in the second string $\mathfrak{T}_2$ (in fact portions of the complete dictionary $W_{m,1}^{[1]}$ with step 1, but with only $(k+1)$ consecutive positions, repeated with step $N-k-1$ between such portions). The sizes of these dictionaries are approximately $k = \lfloor\sqrt{N}\rfloor$ times less than the complete dictionaries $W_{m,1}^{[1]}$, $W_{m,1}^{[2]}$.

We see that for sure at least one complete batch of $k+1$ marks falls inside $S_2 \subset \mathfrak{T}_2$. Then the matching map $M_{S_1,S_2}\colon S_1 \rightleftarrows S_2$ will transfer this batch into a part of S_1 that will include at least one mark in $\mathfrak{T}_1$ (they were taken with step k) provided the corresponding marked position was not deleted by the edit transformation or it will be very close to such a mark if deletions/insertions were situated at positions close to the marked one in $\mathfrak{T}_1$. Then the respective tags in $\mathfrak{T}_1$, $\mathfrak{T}_2$ (entries in $\widetilde{W}_m^{[1]}$, $\widetilde{W}_m^{[2]}$) will approximately match with local error k-density δ or nearby. So using some appropriate method of search for approximately matching entries in $\widetilde{W}_m^{[1]}$, $\widetilde{W}_m^{[2]}$ (for example the method described in [6]) we can find such matching *seed tags* for further *expansion* into the complete approximately matching substrings S_1, S_2 in the way described in more detail below.

Now we give more rigorous description of the process of tag selection, the relation between the parameters N, k, m and δ and prove correctness of the first stage (development of the dictionaries $\widetilde{W}_m^{[1]}$, $\widetilde{W}_m^{[2]}$ with approximately matching tags if there are some approximately matching $S_i \subset \mathfrak{T}_i$).

Theorem 1. *If there are approximately matching substrings $S_1 \subset \mathfrak{T}_1$, $S_2 \subset \mathfrak{T}_2$ of the length N or more with local error v-density δ, $v = \min(m,k)$, then approximately matching tags (substrings of length m) can be found in the dictionary $\widetilde{W}_m^{[1]}$ (developed using constant step k for the sequence $\mathfrak{T}_1$) and in the dictionary $\widetilde{W}_m^{[2]}$ (developed from $\mathfrak{T}_2$ according to the informal description above and the rigorous description below) respectively, with local error m-density δ (i.e. with edit distance at most $d = \delta \cdot m$ between the approximately matching tags) provided that $N \geq 2(k+1) \cdot (1+\delta) + m - 1$.*

Proof. We fix the matching map $M_{S_1,S_2} : S_1 \rightleftarrows S_2$, the prescribed local error v-density δ, some integer k and the tag length m. Next we form the first dictionary $\widetilde{W}_m^{[1]} = W_{m,k}^{[1]}$ as described above. The second dictionary should be formed taking into account the local error m-density δ and possible deletion/insertion errors: we mark batches of $(k+1) \cdot (1+\delta)$ consecutive positions (instead of $k+1$ consecutive positions taken above for simplicity) with $N - 2(k+1) \cdot (1+\delta) - m + 1$ steps between the starting marks of each batch (instead of $N - k - 1$ steps taken above for simplicity).

Then, let us cut off (virtually, for simplicity of the proof) the last $m - 1$ symbols of the substrings S_1, S_2 and look only at the starting positions of the tags of the dictionaries $\widetilde{W}_m^{[i]}$ in such truncated S_i. Obviously at least one complete batch of $(k+1) \cdot (1+\delta)$ marks falls inside this truncated $S_2 \subset \mathfrak{T}_2$ and the matching map $M_{S_1,S_2}: S_1 \rightleftarrows S_2$ will transfer this batch as well as the tags of length m (from the original non-truncated S_2) into a part of S_1 that will include at least one mark in $\mathfrak{T}_1$ (taken with step k) if the corresponding marked position in S_1 was not deleted by the edit transformation. In this favorable case we have the corresponding tag from the first dictionary $\widetilde{W}_m^{[1]}$ that differs from the corresponding tag in $\widetilde{W}_m^{[2]}$ at most by the edit distance $d = \delta \cdot m$ and the statement of the theorem is proved.

In the opposite case when the marked position in $\mathfrak{T}_1$ was deleted by the matching map, we still have a guarantee (cf. the explanation after the Definition 2) that the tag in $\widetilde{W}_m^{[1]}$ starting at the deleted position still has the edit distance not more than $d = \delta \cdot m$ from the tag from $\widetilde{W}_m^{[2]}$ starting at one of the positions that are close to the deleted position after application of the matching map $M_{S_1,S_2}: S_1 \rightleftarrows S_2$. The theorem is proved.

Note that the bound $N \geq 2(k+1) \cdot (1+\delta) + m - 1$ given in the statement of the Theorem is formally much smaller than the bound $N \geq k(k-1) + m - 1$ of Theorem 1 in [15]. This means in fact more freedom in the choice of k in our new algorithm compared to the classical Vernier scale of [15]. In practice one still has a reasonable choice $k = \lfloor \sqrt{N} \rfloor$: if one chooses a larger k then the second dictionary $\widetilde{W}_m^{[2]}$ will be too large; vice versa smaller k yields large $\widetilde{W}_m^{[1]}$ (see discussion in Sect. 5).

In order to find approximately matching tags with bounded error density in the constructed dictionaries one may apply different techniques; in our current implementation the free software [6] was used (cf. Subsect. 4.1). As soon as all approximately matching tags in the dictionaries $\widetilde{W}_m^{[i]}$ are found, the next steps of our algorithm are similar to that in [15]:

- *expand* the found matching tags using their positions in $\mathfrak{T}_i$: consecutively compare the symbols on the right and on the left of the found tags in $\mathfrak{T}_1$ and $\mathfrak{T}_2$ counting non-matching symbols as mutations as well as detected deletions/insertions (one should again use some additional module to detect them) as far as the (local) error density does not exceed the given bound δ, stopping when this error density exceeds δ.
- If the length of the *expanded tag* is at least N in both $\mathfrak{T}_1$ and $\mathfrak{T}_2$, add it to the list of successful *expansions* for further output after all identical tag pairs in $\widetilde{W}_m^{[i]}$ are expanded.

Note that the choice of the parameter m is essential in our approach, it was discussed in [15].

3.2 The Algorithm

Step 1. Given the target length N and δ, choose proper k and m such that $N \geq 2(k+1) \cdot (1+\delta) + m - 1$.

Step 2. If we have two DNA sequences to analyze, develop the dictionaries $\widetilde{W}_m^{[1]}$ and $\widetilde{W}_m^{[2]}$ as described in Subsect. 3.1. Otherwise (for one or more than two DNA sequences if we want to find approximately matching substrings inside symbol sequences as well) develop for each DNA sequence a dictionary with variable step: first take tags of length m starting at positions 1, $k+1$, $2k+1$, ... and then add to this dictionary batches of $(k+1) \cdot (1+\delta)$ consecutive tags of length m, the batches start at positions 1, $N - 2(k+1) \cdot (1+\delta) - m + 1$, $2(N - 2(k+1) \cdot (1+\delta) - m) + 1$, etc. Add the positions of the selected tags into the dictionaries for further expansion.

Step 3. Check whether there are approximately matching entry tags (with the given upper bound on the local error m-density) in the dictionaries using an appropriate external algorithm.

Step 4. *Expand* the found repeated tags (using their positions stored in the dictionaries) as described in Sect. 3.1.

Step 5. List all tagged expansions with their positions in $\mathfrak{T}_i$ and output the list.

More Technical Details. *Finding approximately matching tags on Step 3.* In our current implementation we use the free software [6]. In fact there is a variety of algorithms and techniques to do this approximate matching search; one may vary the algorithm used on this step according to the goals of the particular experiment (cf. Sect. 5).

Expansion strategies on Step 4. If $\delta = 0$, then simple exact matching search and expansion method described in [15] should be applied. If $\delta > 0$, then continue expansion even if the compared symbols around it violate the constraint provided by the local or global error level d; here different strategies for expansion may be useful. For example, one may free error density level for expansion or count, but limit the total number errors globally and/or their global density relative to the total length of the tagged expansions so far (not stopping as soon as the local error m-density will be exceeded).

Treatment of symbols N, W *etc.* In the currently available DNA databases one encounters results with non-exact recognition of nucleotides, they are marked by letters outside of the standard nucleotide alphabet $\aleph = \{A, C, G, T\}$. Several strategies may be applied depending on the problem solved by the researcher; a discussion in [15] gives some guidelines.

4   Experimental Results

To evaluate the efficiency of the proposed algorithm, we have carried out several computational experiments.

To test the most complicated Step 3 (Sect. 3.2) two types of computational experiments have been carried out with the software implementation of the algorithm described in [6]. The first one (Sect. 4.1, Experiment 1) addresses the question of feasibility of the method with some freely distributed software, and on estimation of its run time, for some typical sequences. The main purpose of this experiment is to evaluate the total run time of the program, both for the necessary hash-table implementation and compression, and for a single the query time. The second experiment (Sect. 4.1, Experiment 2) addresses the question towards the feasibility and efficiency of the used software for batch mode queries. All experiments showed reasonable results, proving the feasibility of the method, and usability of the freely distributed software [6].

Section 4.2 shows the results of the expansion Step 4 of our new algorithm.

4.1   Error Tolerant Search

To search the common tags with minor mismatches, we used free software http://code.google.com/p/compact-approximate-string-dictionary/ by Chegrane and Belazzougui [6]. It provides the search of a given string over a given ensemble (a dictionary), with two or less mismatches. A symbol deletion, a symbol insertion, or a symbol mutation are the mismatches allowed by this software. To get the solution, software designers changed the exact hashing for linear probing, eliminated all empty entries in hash-table, and used the direct string comparison. It should be noted that the limitation of two or less mismatches in the search is not a matter of principle; a large number of other approaches without this limitation are known, we simply used the freely available and ready-to-use software [6], other available approaches will remove this limitation, when implemented.

Experiment 1. Single Query Processing. To test the single query time, and the memory necessary to process a query, we used the following data. The first original symbol sequence was the DNA sequence of the XIVth human chromosome. Two dictionaries have been developed over this sequence following the algorithm described in Sects. 3.1 and 3.2; the first used tags of length $m = 50$, and the second used tags of length $m = 30$ (with smaller step k). The first dictionary had 1.7×10^3 entries (tags), and the second one had 1.7×10^6 entries.

At the first run, the query CTGAATCAACCAACCACCTGAAGCTGTCCC was used, with two mutations of symbols. The run time was 80.341 s, and the software returns 34 strings with two mismatches. Neither exact match, nor a single mismatch return has been obtained. The hash-table implementation and its compression took 80.071 s, the exact match search took 0.241 s, a single mismatch search took 0.241 s, and the search with two mismatches took 0.029 s.

At the second run, we used the same string from the dictionary implementing a single insertion, and a single deletion in it. The rum time was 93.458 s, and the software returned two strings differing in two mismatches. Again, neither exact matching, nor a single mismatch return had come back. The hash-table implementation and compression took as long, as 93.022 s, the search for exact match took 0.065 s, and the search with two mismatches took 0.072 s. Slightly more than 3 Gb of memory have been allocated, for the task execution.

For the dictionaries comprising the 50 symbols long string two runs have been executed, also. We used the real tag

CAAGCCACCATACCCAGACATGATGGTCTTTGAAGAAGCGGCCAGTGAAG

with two mutations, to query. The run time was 17.617 s, and the software returned the original string, indicating two mismatches. The hash-table implementation and compression took 17.600 s, the exact matching search took 0.000 s, a single mismatch search took 0.001 s, and two mismatches search took 0.016 s. Similarly, if the string with one deletion and one insertion has been queried, the run time was 19.386 s, and a string has been returned having a mismatch. The software returned 67 strings with two mismatches. The hash-table implementation and compression took 19.341 s, while the exact matching search took 0.000 s, a single mismatch search took 0.004 s, and two mismatches search took 0.041 s. About 0.5 Gb of memory has been allocated, for the task execution.

Experiment 2. Batch Mode Query Processing. We searched for approximately matching tags from the same dictionary with one or two mismatches in batch mode constructing the necessary hash-table implementation and compression and then running a loop of queries for search of approximately matching tags giving consecutively all tags from the dictionary. The queries returned 528 entities with a single mismatch, and 55 302 double-mismatched entities, respectively. The first stage (hash-table implementation and compression) took 17.453 s, while the batch search for single mismatches took 6.906 s, and for double-mismatch search took 497.070 s.

Also, we used two bacterial genomes to develop the dictionaries (these are *Streptomyces bingchenggensis* BCW-1, AC CP002047 in EMBL–bank, 11 936 683 nucleotides long, and *Myxococcus stipitatus* DSM 14675, AC CP004025 in

EMBL–bank, 10 350 586 nucleotides long). Two dictionaries have been developed, both comprising the tags of length 20, for each genome. All tags were labeled with their position number, within the sequence. The dictionaries enlist 119 376 and 109 568 entries, respectively. We used the first dictionary to build the hash-table implementation and batch queries from the second dictionary, looking for approximately matching substrings between the two genomes. For these dictionaries, after batch queries, a few exact matches has been found; 36 single mismatches and 802 approximate matches with double mismatches have been found. Total run time was 36.612 s, including 2.056 s for hash-table implementation and compression, and 34.566 s for the batch search.

4.2 Expansion of the Found Approximately Matching Tags

First we briefly list the results of the Step 4 (Sect. 3.2) for the bacterial genomes *Streptomyces bingchenggensis* and *Myxococcus stipitatus*. After expansion, we got 4 exactly matching substrings of lengths 52, 54 and 67 (two matching pairs for this length). Using all found approximately matching tags and setting the global error density bound to $\delta = 0.25$ we got more than 40 approximately matching substrings, of the length 50 and more, up to the length 1006.

Much more interesting results were obtained in another experiment for the DNA sequence of the XIVth human chromosome. After performing Steps 1 and 2 with $N = 1000$, $k = 30$, $m = 20$ and finding the approximately matching dictionary tags with not more than 2 errors, we got 4,079,978 matching pairs of tags without errors, 5,050,238 matching pairs of tags with 1 error and 17,881,553 matching pairs of tags with 2 errors (mostly ins/dels) thus 27,011,769 approximately matching pairs in total. Expanding the found pairs of tags (*seeds* in the traditional terminology) with the error tolerance $\delta = 0.3$ we got 385,082 expansions of length 1000 or more (the chosen parameter N); among them only 81,204 are in fact different pairs (i.e. many of them were found several times, especially the long ones). It should be noted that we obtain matching *pairs* of substrings of the target length (1000 or more nucleotides) and many substrings are encountered several times (with the given error tolerance), so after the simple clustering of the found pairs w.r.t. the starting points and the lengths of the matching pairs we got approximately 4 times less number of different (approximately) repeating substrings; so the average rate of repeats of a given substring of length 1000 or more is 4. Certainly a number of well-known phenomena were observed: periodic substrings (long runs), a lot of repeats near the end of the XIVth chromosome, etc. Practically all found approximately matching pairs of substrings of length 10,000 and more are in fact long periodic substrings; only two of them were not close to the right end of the chromosome.

Compared to the results obtained for the same XIVth human chromosome by the previous version of the algorithm described in [15] we conclude that the number of approximately matching substrings of the target length 1000 or mire now is dramatically larger: the method of [15] found only 19,946 repeating (exactly matching) tags and only a few hundreds of long repeats (approximate matches with only mutations allowed) of lengths 1000 or more were found after expansion; this is due to the inadequacy of the method of [15] for long repeats with insertions and deletions. The approach presented here as we see from the previous experiment was much more successful in finding long repeats with insertions and deletions permitted. As the Theorem 1 proves, the current algorithm finds *all* such approximately matching long repeats.

The above experiments show that the modified Vernier algorithm proposed in this paper is feasible and can be used in practice.

5   Discussion and Conclusion

In this paper we present basically new algorithm for fast search of approximately matching substrings; the previous version described in [15] addressed the exact matching string search (with minor mismatches). The rarefaction of dictionaries typical of both algorithms brings an advantage in the speed of algorithm execution, since it skips a huge number of comparisons and checks for coincidence of substrings (that is redundant, from the point of view of the search of a common string). The results shown in Sect. 4 demonstrate a 30-fold and even a 100-fold acceleration, on the *seeds* search step. This is a direct consequence of introduction of the parameter N in our algorithm (minimal length of *long repeats* to be found).

An implementation of an approximate search heavily depends on the affordable level of errors to be set up. The method description provided in Sect. 3.2 fixes the parameter that is an average local permissible errors level (that is parameter δ in Sect. 3.1) in substrings under comparison. Formally, one may expect that the parameter may be set up almost arbitrary, and the method still works correctly. Formal correctness yet does not make sense, if δ is great enough: indeed, any two strings may be claimed to be equal, if an arbitrary number of mismatches is allowed. In our paper, we choose the parameter $\delta < 0.5$. Further growth of this parameter may yield any two substrings to be indistinguishable.

We believe that the proposed algorithm with proper choice of parameter (k, m and δ) would be highly effective for usage in large genomic databases. In this respect special attention should be paid to the appropriate choice of the parameters N, k, m and δ of the algorithm proposed. The target length N should be obviously chosen according to the minimal length of long repeats to be found; $m = 20$ or $m = 30$ is usually a good choice depending on the length of genomes to be analyzed. δ is chosen according to the expected error density ($\delta = 0.3$ was good for expansion in our experiments). The parameter k (the step for the dictionary creations) is more delicate; $k = \sqrt{N}$ may be a reasonable choice for analysis of two DNA sequences of approximately the same length, this maximally reduces the dictionary sizes (see the discussion after the end of the proof of Theorem 1). On the contrary, if one would like to improve speed of seeds search in large DNA databases and reduce query time for search of similar genomes to a newly sequenced one, one should consider to choose k only a bit smaller than N, this will result in maximal reduction of the dictionary constructed for the whole DNA database and in fact requires to compose a complete dictionary of m-mers for the DNA sequence sent as the request to the database. More experimentation will be necessary to adapt the parameters of our algorithm to the typical DNA sequences of a given database.

Acknowledgments. This study was supported by a research grant No. 14.Y26.31. 0004 from the Government of the Russian Federation (M.G. Sadovsky) and the grant from Russian Ministry of Education and Science N° 1.8591.2017/VU (S.P. Tsarev).

References

1. Alsmadi, I., Nuser, M.: String matching evaluation methods for DNA comparison. Int. J. Adv. Sci. Technol. **47**, 13–32 (2012)
2. Altschul, S.F., Gish, W., Miller, W., Meyers, E.W., Lipman, D.J.: Basic local alignment search tool. J. Mol. Biol. **215**(3), 403–410 (1990)
3. Altschul, S.F., Madden, T.L., Schaffer, A.A., Zhang, J., Zhang, Z., Miller, W., Lipman, D.J.: Gapped BLAST and PSI-BLAST: a new generation of protein database search programs. NAR **25**(17), 3389–3402 (1997)
4. Bugaenko, N.N., Gorban, A.N., Sadovsky, M.G.: Maximum entropy method in analysis of genetic text and measurement of its information content. Open Syst. Inf. Dyn. **5**(2), 265–278 (1998)
5. Canzar, S., Salzberg, S.L.: Short read mapping: an algorithmic tour. Proc. IEEE **105**(3), 436–458 (2017)
6. Chegrane I., Belazzougui, D.: Simple, compact and robust approximate string dictionary. arXiv:1312.4678v2 [cs.DS] (2014)
7. Gonnet, G.H.: Some string matching problems from Bioinformatics which still need better solutions. J. Dis. Alg. **2**(1), 3–15 (2004)
8. Levenshtein, V.I.: Binary codes capable of correcting deletions, insertions, and reversals. Sov. Phys. Dokl. **10**(8), 707–710 (1966)
9. Maier, D.: The complexity of some problems on substrings and supersequences. J. ACM **25**(2), 322–336 (1978)
10. Marçais, G., Delcher, A.L., Phillippy, A.M., Coston, R., Salzberg, S.L., Zimin, A.: MUMmer4: a fast and versatile genome alignment system. PLoS Comput. Biol. **14**(1), e1005944 (2018)
11. Martin, R.R.: On the computation of edit distance functions. Dis. Math. **338**(2), 291–305 (2015)
12. Pearson, W., Lipman, D.: Improved tools for biological sequence comparison. PNAS **85**, 2444–2448 (1988)
13. Sadovsky, M.G.: Comparison of real frequencies of strings vs. the expected ones reveals the information capacity of macromoleculae. J. Biol. Phys. **29**, 23 (2003)
14. Sadovsky, M.G.: Information capacity of nucleotide sequences and its applications. Bull. Math. Biol. **68**, 156 (2006)
15. Tsarev, S.P., Sadovsky, M.G.: New error tolerant method for search of long repeats in DNA sequences. In: Boton-Fernandez, M., Martin-Vide, C., Santander-Jimenez, S., Vega-Rodríguez, M.A. (eds.) Algorithms for Computational Biology. LNCS, vol. 9702, pp. 171–182. Springer, Cham (2016). https://doi.org/10.1007/978-3-319-38827-4_14
16. https://en.wikipedia.org/wiki/Vernier_scale

Systems Biology and Other Biological Processes

Fixed Parameter Algorithms and Hardness of Approximation Results for the Structural Target Controllability Problem

Eugen Czeizler[1,4]([⊠]), Alexandru Popa[2,3], and Victor Popescu[1]

[1] Department of Computer Science, Åbo Akademi University,
Vesilinnantie 3, 20500 Turku, Finland
{eugen.czeizler,victor.popescu}@abo.fi
[2] Department of Computer Science, University of Bucharest,
Academiei 14, Bucharest, Romania
alexandru.popa@fmi.unibuc.ro
[3] National Institute for Research and Development in Informatics,
Averescu Bd. 8–10, Bucharest, Romania
[4] National Institute for Research and Development of Biological Sciences,
Independentei Bd. 296, Bucharest, Romania

Abstract. Recent research has revealed new applications of network control science within bio-medicine, pharmacology, and medical therapeutics. These new insights and new applications generated in turn a rediscovery of some old, unresolved algorithmic questions, this time with a much stronger motivation for their tackling. One of these questions regards the so-called Structural Target Control optimization problem, known in previous literature also as Structural Output Controllability problem. Given a directed network (graph) and a target subset of nodes, the task is to select a small (or the smallest) set of nodes from which the target can be independently controlled, i.e., it can be driven from any given initial configuration to any desired final one, through a finite sequence of input values. In recent work, this problem has been shown to be NP-hard, and several heuristic algorithms were introduced and analyzed, both on randomly generated networks, and on bio-medical ones. In this paper, we show that the Structural Target Controllability problem is fixed parameter tractable when parameterized by the number of target nodes. We also prove that the problem is hard to approximate at a factor better than $O(\log n)$.

Keywords: Systems biology · Protein interaction networks
Structural network control · Approximation algorithms
Fixed parameter algorithms

© Springer International Publishing AG, part of Springer Nature 2018
J. Jansson et al. (Eds.): AlCoB 2018, LNBI 10849, pp. 103–114, 2018.
https://doi.org/10.1007/978-3-319-91938-6_9

1   Introduction

The network control research field has been investigated for more than 50 years, with some of its algorithmic questions only recently being able to be solved. The general topic is concerned with the optimization of output intervention needed in order to drive a linear, time-invariant, dynamical system from an arbitrary initial state, to a precise final configuration, in finite time. Although many real-life dynamical systems tend not to be linear, most of these systems are known to be well approximated by such dynamics, or could behave as such in specific conditions, such as at their steady state. Although inquiries into this field have been initiated in the 60's and 70's, see e.g. the works in [10,13,18], only in 2011 Liu et al. [14] succeeded to demonstrate that the algorithmic complexity of the full network control optimization problem is actually of a low polynomial complexity, being reduced to computing the maximum matching in a directed graph. The result was received with a lot of interest, and sparked a renewal of the field. Since then, the network control theory and its newly discovered results have been successively applied to the study of control over power grid networks [9], of bio-medical signaling processes [8,11,21], and even the control of social networks [12,14].

Driven by this new insight into the field as well as by its new applications into the current world of Big (or just Large) Data, researchers have realized that full control can sometime be still too expensive. For example, the network control theory has been recently applied in the case of cancer-related bio-medical networks [8,11], with the aim of using known drugs in order to drive the system towards a more favorable state. Thus, researchers aimed at using the protein signaling network in order to drive cancerous cells towards apoptosis, i.e., programmed cell death. However, the full controllability of sparse homogeneous networks, such being many bio-medical networks (e.g. gene signaling networks, metabolic networks, gene regulating networks, etc.) requires a lot of effort, sometimes needing a direct outside control over up to 70% of the initial nodes of the network [11,14]. As in these cases an outside control equivalents to the use of specific drugs, and since these protein networks contain up to 2–3 thousands nodes, a 70% direct outside control would imply an un-viable solution. The key to solving this problem came in the form of a variant of the initial control-theory problem, namely that of target-control. Instead of enforcing the control of the entire network, one would desire to optimize the outside intervention needed to control only a well-specified target, i.e., a subset of the initial network. This proved to be particularly well-fitted with the study of protein signaling networks, as recent research has emphasized the existence of disease-specific essential genes, i.e., disease-specific sets of genes/proteins which, if knocked down, would drive the corresponding cells to apoptosis [1,22,23]. As is the case, new formulations lead to new problems. The Structural Target Control (optimization) problem [3,7] asks to provide an optimum amount of outside intervention in order to drive a linear dynamical system from any initial state to a desired final state of the chosen targets.

Contrary to the full network control case, the Structural Target Controllability problem was proved to be NP-hard [3]. Several heuristic approaches have been implemented and applied to the study of bio-medical networks [3,7,8,11]. However, no detailed analysis of the hardness of approximation have been developed for this problem.

Assuming the widely believed conjecture, that $P \neq NP$, no polynomial time exact algorithms exist for any NP-hard problems. Thus, there are several alternative methods to tackle the difficulty of these problems, such as *approximation algorithms* and *fixed parameter algorithms*. Approximation algorithms run in polynomial time and provide a suboptimal solution. Nevertheless, unlike heuristic algorithms, approximation algorithms guarantee that on every input instance the solution they return is within a certain factor of the optimal solution. For example, a 2-approximation algorithm for a minimization problem, guarantees that on every input, the solution returned is at most twice the size of the optimal solution on that input. However, some problems, such as the one studied in this paper, might not have approximation algorithms with a constant approximation factor, unless $P = NP$. See [20] for a textbook on approximation algorithms.

In practice instances, many problems have parameters that are typically much smaller than the input size. We can exploit the existence of these parameters in order to design faster algorithms for these problems. *Parameterized complexity* [4,6] aims to classify problems according to various parameters that are independent of the size of the input. A fixed parameter algorithm runs in time $f(k)O(n^c)$, where, n is the input size, c is a constant, and k is the size of a parameter (independent of the input size). A problem is termed fixed parameter tractable (FPT) if it has an FPT algorithm.

In this paper we show that the Structural Target Controllability problem is fixed parameter tractable when parameterized by the number of target nodes. Also, if a second parameter is allowed, known in practice to have significantly lower values, the resulted fixed parameter algorithm has a considerably improved complexity. Finally, we also formally prove that the Structural Target Controllability problem is hard to approximate at a factor better than $O(\log n)$.

2 Notation and Preliminaries

A *linear, time invariant dynamical system* (LTIS) is a system

$$\frac{dx(t)}{dt} = Ax(t) \tag{1}$$

where $x(t) = (x_1(t), ..., x_n(t))^T$ is the n-dimensional vector describing the system's state at time t, and $A \in R^{n \times n}$ is the time-invariant *state transition matrix*; the entry $a_{i,j}$ of matrix A describes the weight of the influence of node j over node i. The elements in x are called the *variables* of the system; we denote with X the set of these variables.

The external control over the system is performed through the action of m external *driver nodes*, $u(t) = (u_1(t), ..., u_m(t))^T$. Their influence over the n

variables of the system is described by the time-invariant *input matrix* $B \in R^{n \times m}$; then the LTIS (1), now denoted as (A, B), becomes:

$$\frac{dx(t)}{dt} = Ax(t) + Bu(t) \tag{2}$$

Let $T \subseteq X$, $T = \{t_1, ..., t_k\}$ for some $k \leq n$ be a subset of a particular interest for the variables X, a.k.a., *the target set*. We say that the LTIS (A, B) is T-*target controllable* if for any initial state of the variables in X and any desired numerical setup of the target variables, there exists a time-dependent input vector $u(t) = (u_1(t), ..., u_m(t))^T$ that can drive the system in finite time from its initial state to a state in which the target variables are in the desired final numerical setup. We associate to the k-target set T the characteristic matrix $C_T \in \{0, 1\}^{k \times n}$ where $C_T(i, j) = 1$ iff $i = j$ and $i, j \in T$ (otherwise, $C_T(i, j) = 0$), i.e., C_T is the identity matrix restricted to the subset T. It is known, see e.g. [7], that a system (A, B) is T-target controllable if and only if

$$\text{rank}\mathcal{OC}(A, B, C_T) = |T| \tag{3}$$

where the matrix $\mathcal{OC}(A, B, C_T) := [C_T B \mid C_T A B \mid C_T A^2 B \mid ... \mid C_T A^{n-1} B]$ is called the *controllability matrix*.

In the particular case when the target is the entire n variable set X, the above condition translates to the well known Kalman's condition for full controllability [10], i.e., an LTIS (A, B) is (fully) controllable if and only if $\text{rank}[B \mid AB \mid A^2 B \mid ... \mid A^{n-1} B] = n$.

The notion of target controllability and the focus of imposing a controlling effect only on a subset of the variables of the system, has been introduced and studied only recently, see e.g., [3,7,8,11]. However, this notion can be seen as a special case of output controllability, a topic which received considerate attention in the 80's and 90's, see. e.g. the works of Poljak and Murota [15–17].

Although the control methodology seem to be very dependent on the numerical setup of the dynamical system of our choice, i.e., the numerical setup of the associated transition matrix A, it turns out that this is not the case. We say that an LTIS (A, B) is T-*structurally target controllable* (with respect to a given size-k target set T) if there exists a time-dependent input vector $u(t) = (u_1(t), ..., u_m(t))^T$ and a numerical setup for the non-zero values within the matrices A and B, that can drive the state of the target nodes to any desired numerical setup in finite time. A deep result of [13,18] shows that a system is structurally target controllable if and only if it is target controllable for all structurally equivalent matrices A and B, except a so-called "thin" set of matrices; we say that two matrices are *structurally equivalent* iff they differ only on their non-zero values.[1] Thus, the existence of "a good choice" for the numerical parameters in A and B is (almost) equivalent to picking up any numerical values for these parameters. According to Eq. (3) above, for a k-sized target T, a system

---

[1] It is beyond the goal of this paper to define the topological notion of thin sets; we only give here the intuition that such sets consist of isolated cases that may be easily replaced with nearby favorable cases.

(A, B) is structurally T-target controllable if and only if there exist values for the non-zero entries in A, B such that rank$\mathcal{OC}(A, B, C_T) = |T| = k$.

It is known, see e.g. [16,17], that the structural controllability problem has a counterpart formulation in terms of graphs/networks. Given an LTIS (A, B), we associate to it the graph $G_{(A,B)} = (V, E)$ where the n variables of the system $\{x_1, ..., x_n\}$ and the size-m external controller $\{u_1, ..., u_m\}$ are the nodes of the graphs, while directed edges correspond to the non-zero values in the state transition matrix and input matrix, respectively. That is, there exists a directed edge from the node corresponding to variable x_i to the node corresponding to x_j if and only if $A(x_j, x_i) \neq 0$.[2] Similarly, there exists a directed edge from u_i to x_j if and only if $B(x_j, u_i) \neq 0$. The nodes $\{u_1, ..., u_m\}$ are called *driver nodes*, while the nodes x_j such that there exists i with $B(x_j, u_i) \neq 0$ are called the *driven nodes* of the network. In the literature, the driver and the driven nodes are sometimes known as *input* and *controlled* nodes [7,14]. To a rough understanding, the difference between driver and driven nodes is as follows. The set of driver nodes is describing the complexity of an outside controller, assuming this controller can interact/influence independently several well specified nodes of the network. Meanwhile, the set of driven nodes provides exactly the exact collection of network nodes that are used in order to ultimately control the entire set of targets. From an algebraic perspective, the number of driver nodes is given by the number of (nonzero) columns of the control matrix B, while the number of driven nodes is given by the number of nonzero rows of B. It was shown in [3] that from a practical perspective, it is more meaningful to analyze the controllability optimization problem from the point of view of minimizing the number of driven nodes. This is why in this research we focus on this particular formulation of the optimization problem. Thus, we impose that each driver node is connected to exactly one driven node, i.e., the input matrix B contains exactly one non-zero element on each column.

Given an LTIS (A, B) and its associated graph $G_{(A,B)} = (V, E)$, the n variables of the system are (all) structurally controllable from the m-sized input controller u (and control matrix B) if and only if we can select a set of n directed paths from driver nodes as starting points (we denote this set as $\mathcal{U}$) to each of the network nodes, as ending points, such that no two paths would intersect at the same distance d from their end points. The above formulation is closely related to the concepts of *linking* and *dynamic graph* as investigated in [16,17]. In case of the target controllability problem, for a given target set $T = \{t_1, t_2, ..., t_k\} \subseteq X$, the above graph formulation is naturally adjusted as follows. We introduce k new *output nodes* $\mathcal{C}_T = \{c_1, c_2, ..., c_k\}$ (also denoted as $\mathcal{C}$ when clear from the context) and edges (t_i, c_i), for all $1 \leq i \leq k$. Note that the output matrix C_T describes exactly the above wiring. Now, the objective becomes to find a path family containing k directed paths, connecting all the driver nodes (as start-points) to the output nodes (as end-points), such that no two paths would intersect at the same distance d from their end-points. In contrast to the case of full control, the graph condition is only necessary for

---

[2] We implicitly interchange the usage of x_i and i for matrix indices.

target control, but not sufficient [16]. However, as investigated in [3], it is only in very restrictive cases where the existence of such a path family would not translate into the algebraic definition of structural control. Thus, from all practical purposes, one can equate the algorithmic process of finding such a family of k directed path to verifying that the system is structural target controllable.

We define the notion of optimization for structural target controllability in case of LTIS as follows:

Definition 1. The Structural Target Control (Optimization) problem (in short STC):

Input: The size-n variable set X, the associate transition matrix A of size $n \times n$, and a size-k target subset $T \subseteq X$, with $k \leq n$.

Output: Matrix B of size $n \times m$ such that

1. *every column of B contains exactly one non-zero value,*
2. *$SrankOC(A, B, C) = k$,*
3. *m (i.e., the number of columns of B) is minimum among all feasible matrices.*

3   Fixed Parameter Algorithm

In this section we prove that the STC problem is fixed parameter tractable, parameterized by some of the secondary variables of our problem. First, we show that one parameter, namely the number of target nodes $|T| = k$, suffice in generating such a fixed parameter algorithm. On the other hand, from the practical instances from where this problem was generated, namely the targeted control of human protein signaling networks in cancer, we identify several other variables of this problem which are known to have significantly lower numerical values, i.e., one or even two orders of magnitude lower than the total number of input nodes. Thus, we will involve these parameters in order to generate some lower complexities for the structural target control optimization problem.

3.1   A One-Parameter STC Algorithm

Informally, our algorithm carries out the following steps. First we compute for each vertex v in the input graph, all the possible subsets of T that v can control. Since $T = k$, there can be at most 2^k such subsets for each node v. Then, we enumerate over all possible subsets of 2^T (notice that there are precisely 2^{2^k} of such subsets). For each such subset of $\mathcal{D} \subseteq 2^T$ we check if there exists a collection of $|\mathcal{D}|$ nodes such that each node controls precisely one set in $\mathcal{D}$. If so, we solve exactly the set cover instance $(\mathcal{D}, T)$ and store the solution if it is better than the previously found solutions (i.e., controls the target nodes with less nodes than the previous solutions). Algorithm 1 describes our procedure in detail.

Theorem 2. *Given a graph $G = (V, E)$ and a target set $T \subseteq V$ with $|T| = k$, Algorithm 1 solves the STC problem in time $O(f(k)p(n))$. Thus, the STC problem is fixed parameter tractable.*

Algorithm 1. An FPT algorithm for the STC problem

Input: An undirected graph $G = (V, E)$ and a set of nodes $T \subseteq V$, $|T| = k$
Output: A set of nodes $S \subseteq V$ of minimum cardinality that controls T.

1. For every node $v \in V$, compute all possible sets of target nodes that v can control in the same time $C_v \subseteq 2^T$.
2. $OPT := \infty$, $S = \emptyset$
3. For every $\mathcal{D} \subseteq 2^T$ do:
   (a) Let $\mathcal{D} = \{D_1, D_2, \ldots, D_\ell\}$. If there exists nodes $v_1, v_2, \ldots, v_\ell$ such that $C_{v_1} = D_1, C_{v_2} = D_2, \ldots, C_{v_\ell} = D_\ell$, then:
   (b) Solve exactly the set cover problem on instance $(\mathcal{D}, T)$. Let $\mathcal{D}' = \{D_{u_1}, D_{u_2}, \ldots, D_{u_x}\}$ be the sets in the optimal set cover. If $x < OPT$, then $OPT := x$ and $S := \{u_1, u_2, \ldots, u_x\}$

return S

Proof. We present in more detail and analyze the running time of each step of Algorithm 1.

Step 1. For each node $v \in V$ we compute and store as follows all the sets of nodes in T that v can simultaneously control. First, we show how to decide if a node $v \in V$ covers a given subset of nodes $T' \subseteq T$ in polynomial time in $|V|$. Given a set of vertices $X \subseteq V$, let $N(X)$ be the neighborhood of X, that is $N(X) = \{v \in V : \exists a \in X \text{ s.t. } (v, a) \in E\}$. Define the following graph $G_{v,T'} = (V', E')$ where:

1. Let $T_0 = T$ and $T_{i+1} = N(T_i)$, $\forall 0 \leq i < n$. The vertex set V' of the graph $G_{v,T'}$ is the *multiset* consisting of all the sets T_i plus two other vertices $\{s, t\}$. We refer to a vertex $p \in V$ that is in the set T_i as p^i. Notice that a vertex p cannot appear twice in a set T_i.
2. In the edge set E' of the graph $G_{v,T'}$ we add an edge (a^{i+1}, b^i) if $(a, b) \in E$. Moreover, we add an edge between (s, v^i), if $v^i \in T_i$. Finally we add an edge (a, t), $\forall a \in T'$

The vertex v can control simultaneously the nodes in the set V if and only if there exists k-vertex disjoint paths from s to t. Observe that graph $G_{v,T'}$ was constructed such that any two vertex disjoint paths from s to t in $G_{v,T'}$ correspond to paths in G from v to a vertex in T' that do not intersect at the same distance from the vertices in T'. The k vertex disjoint paths problem between two vertices is solvable in time $O(k(n + m))$ on a graph with n vertices and m edges [2]. Thus, since $G_{v,T'}$ has at most $|V|^2$ vertices and $|V|^3$ edges, to find k disjoint paths between s and t, takes time at most $k|V|^3$.

Then, to complete step 1, we repeat the procedure described above for every vertex $v \in V$ and any subset $T' \subseteq T$. Since there are 2^k subsets of T, the total running time of step 1 is $O(2^k k |V|^4)$.

Step 2(a)

Since any set C_v has at most 2^k elements and any set $\mathcal{D}$ has at most 2^{2^k} elements, step 2(a) of the algorithm is solved in time $O(n2^{2^k})$: for every node $v \in V$ we simply search each element of C_v in $\mathcal{D}$.

Step 2(b)

Notice that since the number of sets in the set cover instance is bounded by 2^{2^k} and the number of elements is k, then we can solve the set cover in $O(2^{2^{2^k}})$ time by a simple brute force algorithm that chooses all the possible subsets of $\mathcal{D}$ and verifies if such a subset covers T.

Since Steps 2(a) and 2(b) are executed 2^{2^k} times, the total running time of Step 2 is $O(2^{2^{2^k}})$.

Thus, the overall running time of Algorithm 1 is $O(k2^k|V|^4 + 2^{2^{2^k}})$

3.2   Towards Tractable Full Search Algorithms by Multiple-Parameterization

In the following, we present a fixed parameter tractable algorithm for STC whose runtime complexity is exponential in the parameters k and p, corresponding to the size of the target set T and the maximal length of the controlling path from a driver to a target node, respectively, and low polynomial in n, the total number of nodes in the network. The algorithm is a full search expansion of a Greedy approach first reported in [7] and later analyzed and improved in [3,8,11].

Algorithm 2. An FPT algorithm for the STC problem parametrized by k, the size of the target set and p, the maximal length of the controlling path

Input: A directed graph $G = (V, E)$, a set of nodes $T \subseteq V$, $|T| = k$, and an integer p.
Output: A set of nodes $U \subseteq V$ of minimum cardinality that controls T.

1. We create a new graph $G' = (V', E')$. For determining V' we add to V a number of k nodes (denoted $u_1, u_2, ..., u_k$) and for E' we add to E a number of k edges, such that the edge $(u_i \rightarrow t_i) \in E'$, $\forall i = \overline{1, k}$.
2. We set $S_{best} = T$, $|S_{best}| = k$ and $S = \emptyset$.
3. We apply the iterative algorithm Control (Algorithm 3) for $(G' = (V', E'), i = 1, T_0 = T, p, S)$.

return S_{best}

Theorem 3. *Given a graph $G = (V, E)$ and a target set $T \subseteq V$ with $|T| = k$ and $|V| = n$, Algorithm 2 solves the Target Controllability Problem in time $O(kn \cdot (\frac{e(n+k)}{k})^{kp})$.[3] By further assuming a ratio of $1/10$ between the size of the target set vs. the total nodes, we obtain an approximate time complexity $O((11e)^{kp} \times n)$.*

---

[3] We use the following upper bound for the binomial coefficient $\binom{n+k}{k} \leq (\frac{e(n+k)}{k})^k$.

Algorithm 3. The iterative function Control called in the main program

Input: A directed graph $G = (V, E)$, an integer i- the current level in the linking graph, two sets of nodes S- the current solution (incomplete if $i < p$) and T_{i-1}- the current target in the ith level of the linking graph, and an integer p- the maximum expansion of the linking graph.

Output: The set T_i which is the target in the $(i + 1)^{\text{th}}$ level of the linking graph and an update of S, the current solution for the driven set. If $i = p$, a possible update of the S_{best} solution.

1. We build a bipartite graph G_i with the nodes in V on the left side (denoted T_i), and the nodes in T_{i-1} on the right side. We add to G_i all of the edges in E that have the source node in T_i and the destination node in T_{i-1}.
2. We compute/enumerate all maximal matchings in the graph G_i between the nodes in T_i and the nodes in T_{i-1}.
3. For each maximal matching, do:
   (a) We remove from T_i all of the nodes left unmatched. We add all unmatched nodes from T_{i-1} to S, if they are not already there. If they are, then they are left unchanged.
   (b) (Optionally, to speed up the search, we check if $|S| \geq |S_{best}|$, and if so we backtrack)
   (c) If $i = p$, we add to S all of the nodes in T_i. If $|S| < S_{best}$, then $S_{best} \leftarrow S$.
   (d) If $i \neq p$, we repeat again the iterative algorithm for $(G' = (V', E'), i+1, T_i, p, S)$

We omit the proof due to space limitations.

Proof. In the following, we present in more details and analyze the running time of each step of the Algorithm 2 and of its Control sub-function, i.e., Algorithm 3.

The final controlling set, S_{best}, can be updated only after p nested applications of the iterative Control algorithm. In each of these p nested steps, we need to generate a bipartite graph, compute/enumerate all possible maximal matchings, and form the set S, which will then be fed into the next application of the iterative function Control. While the construction of the bipartite graph can be done in $O(kn)$, enumerating all its maximal matchings requires $O(n)$ per maximal matching, see e.g. [19]. In the worst case scenario, when we are dealing with a complete graph G, all of the intermediary bipartite graphs G_i will also be complete. Thus, in each case, the number of edges will be bounded by $k \cdot (n + k)$ (since we have $|V'| = n + k$ nodes on the left side, and $|T_i| \leq k$ nodes on the right side) while the number of maximal matchings will be upper bounded by $\binom{n+k}{k}$. Therefore, the overall time complexity can be upper bounded by $O(kn + \binom{n+k}{k} \cdot (1 kn + \binom{n+k}{k} \cdot (2...p \text{ times...})_2)_1)$, i.e., $O(\binom{n+k}{k}^p \cdot kn)$. As $\binom{n+k}{k} \leq (\frac{e(n+k)}{k})^k$, we get that the running time of the algorithm can be upper bounded by $O(kn \cdot (\frac{e(n+k)}{k})^{kp})$.

4   Hardness of Approximation

In this section we show that the Structural Target Controllability (optimization) problem cannot be approximated within a factor of $(1 - \epsilon) \ln k, \forall \epsilon > 0$ where k is the number of nodes in the target set T. We prove this via an approximation preserving reduction from the Set Cover problem, which is known to be hard to approximate by Feige [5].

Definition 4 (Set Cover). *Given an universe of elements $\mathcal{U} = \{u_1, u_2, \ldots, u_k\}$ and a family consisting of n subsets of $\mathcal{U}$, $\mathcal{S} = \{S_1, S_2, \ldots, S_n\}$, find the smallest sub-collection $\mathcal{S}' \subseteq \mathcal{S}$ such that the union of all the sets $\mathcal{S}'$ is $\mathcal{U}$.*

Theorem 5. *Unless $NP \subseteq DTIME(n^{\log \log n})$, the STC problem cannot be approximated within a factor of $(1 - \epsilon) \ln n, \forall \epsilon > 0$.*

Proof. Given an instance of the Set Cover problem, i.e., a set $\mathcal{U} = \{u_1, u_2, \ldots, u_k\}$ with k elements and n sets $S_1, S_2, \ldots, S_n \subseteq \mathcal{U}$, we construct the following instance of the STC problem.

1. Add a vertex $s_i \in V$ corresponding to each set S_i in the Set Cover instance.
2. Add a vertex $t_i \in V$ corresponding to each element u_i in the set $\mathcal{U}$.
3. For each S_i add $q_i = |S_i|(|S_i| - 1)/2$ auxiliary vertices in V. We term these vertices $a_1^i, a_2^i, a_3^i, \ldots, a_{q_i}^i$
4. The target set T consists of all the nodes $t_i \in V$
5. For each set S_i of the set cover instance we construct $|S_i|$ paths of length $2, 3, 4 \ldots |S_i| + 1$ as follows. Let $S_i = \{u_1, u_2, \ldots u_{|S_i|}\}$. Then we construct the paths: $\{s_i, u_1\}, \{s_i, a_1^i, t_1\}, \{s_i, a_2^i, a_3^i, t_2\}, \ldots, \{s_i, a_{q_i - |S_i| + 1}^i, a_{q_i - 1}^i, \ldots, a_{q_i}^i, t_{|S_i|}\}$

We show now that the Set Cover instance has a solution with x sets if and only if the target set of nodes T can be controlled with x driver nodes. Thus, the existence of an approximation algorithm of $(1 - \epsilon) \ln n$, for some $\epsilon > 0$, implies the existence of an approximation algorithm with the same factor for the Set Cover problem (which leads to a contradiction).

Given a Set Cover with x sets $S_{i_1}, S_{i_2}, \ldots S_{i_x}$, then the driver nodes $s_{i_1}, s_{i_2}, \ldots s_{i_x}$ control all the target nodes since each s_{i_j} controls precisely the target nodes corresponding to the elements in S_{i_j}. This holds since each path from the node s_{i_j} to vertices in T has a different length.

Conversely, given a set of x driver nodes that control all the target nodes we reconstruct a valid Set Cover with x sets, by choosing the sets corresponding to the driver nodes. Thus, the theorem follows.

5   Conclusions and Future Work

Network Science has been proven to be highly relevant within the current developments of medicine and of personalized therapeutics. Within this field, structural network control is a powerful and efficient tool for steering the involved bio-medical systems towards desirable configurations. Thus, the algorithmic optimization problems studied in this manuscript are relevant for the computational bio-medicine community, as highly optimized solutions have a significant chance of translating into efficient therapeutics. Although the Structural Target Control (Optimization) problem has been proven to be NP-hard in its general case and can not even be approximated within a constant factor, and although it is a known fact that bio-medical networks are rather large, containing thousands of nodes and (tens of thousands of) interactions, in practice, several of the involved parameters can still be considerably bounded to significantly lower values. In this research we took advantage of these insights in order to provide two optimization algorithms which remain of low polynomial complexity with regards to the size of the network, and are exponential only in those chosen parameters.

However, Structural Controllability is only one of network science methods which can be used in order to influence the dynamics of these systems. Other methods, such as the Minimum Dominating Set (MDS) approach, or the Target Reachability approach come with new challenges, but also with several advantages. Thus, the optimization and approximation of these algorithms is of a similar practical importance, and worth of detailed investigation and analysis.

Acknowledgments. This work was supported by the Academy of Finland through grant 272451, by the Finnish Funding Agency for Innovation through grant 1758/31/2016, by the Romanian National Authority for Scientific Research and Innovation, through the POC grant P 37 257, and by the (Romanian) Ministry of Research and Innovation through the Institutional Research Programme PN 1819, project PN 1819-01-01.

References

1. Blomen, V.A., et al.: Gene essentiality and synthetic lethality in haploid human cells. Science **350**(6264), 1092–1096 (2015). https://doi.org/10.1126/science.aac7557, http://science.sciencemag.org/content/350/6264/1092
2. Bondy, A., Murty, U.: Graph Theory. Springer, Heidelberg (2011). https://books.google.ro/books?id=HuDFMwZOwcsC
3. Czeizler, E., et al.: Structural target controllability of linear networks. IEEE/ACM Trans. Comput. Biol. Bioinform. **PP**(99), 1 (2018). https://doi.org/10.1109/TCBB.2018.2797271
4. Downey, R.G., Fellows, M.R.: Parameterized Complexity. Springer, Heidelberg (2012)
5. Feige, U.: A threshold of ln n for approximating set cover. J. ACM **45**(4), 634–652 (1998). https://doi.org/10.1145/285055.285059
6. Flum, J., Grohe, M.: Parameterized Complexity Theory. Texts in Theoretical Computer Science. An EATCS Series. Springer, Heidelberg (2006). https://doi.org/10.1007/3-540-29953-X
7. Gao, J., Liu, Y.Y., D'Souza, R.M., Barabási, A.L.: Target control of complex networks. Nat. Commun. **5**, 5415 (2014). https://doi.org/10.1038/ncomms6415
8. Guo, W.F., et al.: A novel algorithm for finding optimal driver nodes to target control complex networks and its applications for drug targets identification. BMC Genomics **19**(1), 924 (2018). https://doi.org/10.1186/s12864-017-4332-z

9. Hines, P., Blumsack, S., Sanchez, E.C., Barrows, C.: The topological and electrical structure of power grids. In: 2010 43rd Hawaii International Conference on System Sciences, pp. 1–10, January 2010. https://doi.org/10.1109/HICSS.2010.398

10. Kalman, R.E.: Mathematical description of linear dynamical systems. J. Soc. Ind. Appl. Math. Ser. A Control **1**(2), 152–192 (1963). https://doi.org/10.1137/0301010

11. Kanhaiya, K., Czeizler, E., Gratie, C., Petre, I.: Controlling directed protein interaction networks in cancer. Sci. Rep. **7**(1), 10327 (2017). https://doi.org/10.1038/s41598-017-10491-y

12. Li, A., et al.: The fundamental advantages of temporal networks. Science **358**(6366), 1042–1046 (2017). https://doi.org/10.1126/science.aai7488, http://science.sciencemag.org/content/358/6366/1042

13. Lin, C.: Structural controllability. IEEE Trans. Autom. Control **19**, 201–208 (1974)

14. Liu, Y.Y., Slotine, J.J., Barabási, A.L.: Controllability of complex networks. Nature **473**(7346), 167–173 (2011). https://doi.org/10.1038/nature10011

15. Murota, K.: Systems Analysis by Graphs and Matroids: Structural Solvability and Controllability. Algorithms and Combinatorics. Springer, Heidelberg (1987). https://books.google.fi/books?id=DkHvAAAAMAAJ

16. Murota, K., Poljak, S.: Note on a graph-theoretic criterion for structural output controllability. KAM series, discrete mathematics and combinatorics, operations research, mathematical linguistics, Department of Applied Mathematics, Charles University (1989). https://books.google.fi/books?id=5RrPHAAACAAJ

17. Poljak, S.: On the generic dimension of controllable subspaces. IEEE Trans. Autom. Control **35**(3), 367–369 (1990). https://doi.org/10.1109/9.50361

18. Shields, R., Pearson, J.: Structural controllability of multiinput linear systems. IEEE Trans. Autom. Control **21**(2), 203–212 (1976). https://doi.org/10.1109/TAC.1976.1101198

19. Uno, T.: Algorithms for enumerating all perfect, maximum and maximal matchings in bipartite graphs. In: Leong, H.W., Imai, H., Jain, S. (eds.) ISAAC 1997. LNCS, vol. 1350, pp. 92–101. Springer, Heidelberg (1997). https://doi.org/10.1007/3-540-63890-3_11

20. Vazirani, V.V.: Approximation Algorithms. Springer, New York, Inc. (2001)

21. Vinayagam, A., et al.: Controllability analysis of the directed human protein interaction network identifies disease genes and drug targets. Proc. Natl. Acad. Sci. **113**(18), 4976–4981 (2016). https://doi.org/10.1073/pnas.1603992113

22. Wang, T., et al.: Identification and characterization of essential genes in the human genome. Science **350**(6264), 1096–1101 (2015). https://doi.org/10.1126/science.aac7041. http://science.sciencemag.org/content/350/6264/1096

23. Zhan, T., Boutros, M.: Towards a compendium of essential genes - from model organisms to synthetic lethality in cancer cells. Crit. Rev. Biochem. Mol. Biol. **51**(2), 74–85 (2016). https://doi.org/10.3109/10409238.2015.1117053, PMID: 26627871

Symbolic Detection of Steady States of Autonomous Differential Biological Systems by Transformation into Block Triangular Form

Chenqi Mou[✉] [iD]

Beijing Advanced Innovation Center for Big Data
and Brain Computing/LMIB – School of Mathematics and Systems Science,
Beihang University, Beijing 100191, China
chenqi.mou@buaa.edu.cn

Abstract. In this paper we propose a method for transforming a square polynomial set into block triangular form by using Tarjan's algorithm. The proposed method is then applied to symbolic detection of steady states of autonomous differential biological systems which are usually sparse systems with a large number of loosely coupling variables. Two biological systems of 12 and 43 variables respectively are studied to illustrate the effectiveness of the proposed method.

Keywords: Systems biology · Differential biological system
Steady state · Block triangular form · Sparsity

1 Introduction

Many biological systems such as chemical reaction networks and reconstructed signal pathways can be modeled mathematically by dynamic systems (see, e.g. [1,15,16]). Algebraic approaches have been successfully applied to the detection and stability analysis of equilibria of biological dynamic systems for both continuous and discrete systems [17,19,24,28] and the analysis of bifurcations and limit cycles for continuous systems [10,23,25]. In these applications of algebraic methods to such analysis of biological dynamic systems, the problems of interests for the biological systems are usually first reduced into algebraic problems like finding real solutions of polynomial systems (or semi-algebraic systems) and finding the ranges of the parameters for which the algebraic or semi-algebraic systems have prescribed numbers of real solutions, and then symbolic methods, including but not limited to those for solving polynomial systems like Gröbner bases [4,11] and triangular decomposition [2,27], for quantifier elimination like

This work was partially supported by the National Natural Science Foundation of China (NSFC 11401018 and 11771034).

CAD and partial CAD [6,7], and for real root isolation and real solution classification [29], are called to solve the reduced algebraic problems.

In this paper we are interested in the symbolic detection of steady states of autonomous differential biological systems which can be easily reduce to solving polynomial systems symbolically. Compared with numeric methods for solving polynomial systems, symbolic methods produce rigorous and reliable solutions which are convenient for later manipulations (for example the stability analysis after the steady states of dynamic systems are computed), but the applications of symbolic methods are also limited by their relatively low computational efficiency compared with numeric methods. To overcome this difficulty, efficient specialized algorithms for symbolically solving polynomial systems have been proposed for sparse systems [12,14], systems with symmetry [13], systems with special graph structures (e.g., chordal graphs) [5,22].

Biological dynamic systems usually involve a large number of interacting variables governed by differential equations. Typically finding all the real solutions of large polynomial systems like these ones with symbolic methods are difficult computationally. However, in many biological systems like chemical reaction networks or reconstructed signal pathways [3,21], the connections between the variables are loose, resulting in sparse polynomial systems with respect to the variables. In particular, the structures of real solutions of sparse polynomial systems in chemical reactions are studied in [16] by using graph theory.

In this paper we make use of the sparsity of polynomial systems arising from autonomous differential biological systems by transforming them into block triangular form and then solving the resultant polynomial systems blockwise. This strategy can be viewed as multivariate generalization of similar techniques for solving sparse linear systems [9,26] and it is effective for studying autonomous differential biological systems for they are usually large and sparse as mentioned above.

The outline of this paper is as follows. After presenting necessary notions and notations for polynomial sets in block triangular form and Tarjan's algorithm for sparse linear systems in Sect. 2, we describe the method we proposed for symbolic detection of autonomous differential biological systems by blockwise solving polynomial sets obtained after transformation into block triangular form by using Tarjan's algorithm in Sect. 3. In Sect. 4 two biological systems of 12 and 43 variables respectively are studied with the proposed method to illustrate its effectiveness, and the paper ends with some concluding remarks in Sect. 5.

2   Preliminaries

Let $\mathbb{K}[x_1, \ldots, x_n]$ be a multivariate polynomial ring, where $\mathbb{K}$ is a ground field and $x_1, \ldots, x_n$ are the variables ordered as $x_1 < x_2 < \cdots < x_n$. We write $\boldsymbol{x}$ for $\{x_1, \ldots, x_n\}$ and $\mathbb{K}[\boldsymbol{x}]$ for $\mathbb{K}[x_1, \ldots, x_n]$ respectively for simplicity. These notations will be fixed hereafter.

2.1   Triangular Sets and Polynomial Sets in Block Triangular Form

Triangular Sets. For an arbitrary polynomial $P \in \mathbb{K}[x]$, the greatest variable which effectively appears in P is called the *leading variable* of P and denoted by $\mathrm{lv}(F)$. Suppose that $\mathrm{lv}(P) = x_k$. Then P can be written as $P = I x_k^d + R$ with $I \in \mathbb{K}[x_1, \ldots, x_{k-1}]$, $R \in \mathbb{K}[x_1, \ldots, x_k]$, and $\deg(R, x_k) < d$. The polynomial I here is called the *initial* of P and denoted by $\mathrm{ini}(P)$.

Definition 1. An ordered set of non-constant polynomials $\mathcal{T} = [T_1, \ldots, T_r] \subset \mathbb{K}[x]$ is called a *triangular set* if $\mathrm{lv}(T_1) < \cdots < \mathrm{lv}(T_r)$.

The special structure of triangular sets makes them easy to solve by successively solving univariate polynomials after substitution of computed partial solutions. For two polynomial sets $\mathcal{P}, \mathcal{Q} \subset \mathbb{K}[x]$, the set of common zeros of $\mathcal{P}$ is denoted by $\mathsf{Z}(\mathcal{P})$, and $\mathsf{Z}(\mathcal{P}/\mathcal{Q}) := \mathsf{Z}(\mathcal{P}) \setminus \mathsf{Z}(\prod_{Q \in \mathcal{Q}} Q)$.

Definition 2. Let $\mathcal{P} \subset \mathbb{K}[x]$ be a polynomial set. Then a finite number of triangular sets $\mathcal{T}_1, \ldots, \mathcal{T}_r \subset \mathbb{K}[x]$ are called a *triangular decomposition* of $\mathcal{P}$ if the zero relationship $\mathsf{Z}(\mathcal{P}) = \cup_{i=1}^r \mathsf{Z}(\mathcal{T}_i/\mathrm{ini}(\mathcal{T}_i))$ holds, where $\mathrm{ini}(\mathcal{T}_i) := \{\mathrm{ini}(T) : T \in \mathcal{T}_i\}$.

Like the method based on Gröbner bases, triangular decomposition is a standard symbolic method for solving polynomial systems. The process of computing a triangular decomposition $\mathcal{T}_1, \ldots, \mathcal{T}_r$ of $\mathcal{P}$, which can be viewed as multivariate generalization of Gaussian elimination for reducing a square matrix to echelon form, reduces the problem of finding all the solutions of $\mathcal{P} = 0$ to solving $\mathcal{T}_i = 0$ for $i = 1, \ldots, r$ where each $\mathcal{T}_i$ is a triangular set and thus easy to solve.

The elimination properties of triangular sets and Gröbner bases permit constructive procedures for solving polynomial systems in a generalized way of solving linear systems in echelon form. We want to mention that computationally the difficulty of solving polynomial systems with these symbolic methods is sensitive to the numbers of variables of the systems, for example in the worst case the complexity of computing Gröbner bases is doubly exponential in the number of variables [20].

Block Triangular Form. Let $\mathcal{P} \subset \mathbb{K}[x]$ be a polynomial set and $X = [x_1, \ldots, x_r]$ be a partition of the variable set x, where each x_i is a subset of $x = \{x_1, \ldots, x_n\}$ for $i = 1, \ldots, r$. Then $\mathcal{P}$ is said to be in *block triangular form* with respect to X if $\mathcal{P}$ can be written in the form

$$\mathcal{P}_1(x_1), \mathcal{P}_2(x_1, x_2), \ldots, \mathcal{P}_r(x_1, \ldots, x_r), \tag{1}$$

where $\mathcal{P}_i \subset \mathbb{K}[x]$ for $i = 1, \ldots, r$ and $\{\mathcal{P}_1, \mathcal{P}_2, \ldots, \mathcal{P}_r\}$ forms a partition of the polynomial set $\mathcal{P}$. In the case when $\#\mathcal{P}_i = 1$ for $i = 1, \ldots, n$, the polynomial set $\mathcal{P}$ is said to be in *strict triangular form*. By comparisons to the definition of triangular sets (Definition 1), one can find that when in strict triangular form, $\mathcal{P}$ is a triangular set with respect to any variable ordering compatible with $x_1 < x_2 < \cdots < x_r$.

When a polynomial set $\mathcal{P} \subset \mathbb{K}[x]$ is in block triangular form (1), solving the polynomial equation system $\mathcal{P} = 0$ is reduced to successively solving each $\mathcal{P}_i$ with respect to x_i after the substitution of solutions $\overline{x}_1, \ldots, \overline{x}_{i-1}$ of $\mathcal{P}_1, \ldots, \mathcal{P}_{i-1}$ respectively for $i = 1, \ldots, r$. Therefore the concept of polynomial sets in block triangular form is generalization of triangular sets and also multivariate generalization of matrices in echelon form in Gaussian elimination. As in the previous discussions, the complexity of solving a polynomial system is heavily dependent on the number of its variables, this kind of reduction into solving subsystems is effective when r, the number of sets in the partition, is relatively large.

2.2   Tarjan's Algorithm for Transforming Matrices into Block Triangular Form

A square matrix A over $\mathbb{K}$ is said to be in *block triangular form* if A can be written as

$$
A = \begin{pmatrix} A_{1,1} & & & \\ A_{2,1} & A_{2,2} & & \\ \vdots & \vdots & \vdots & \\ A_{r,1} & A_{r,2} & \cdots & A_{r,r} \end{pmatrix},
$$

where $A_{i,j}$ is a square submatrix for $i = 1, \ldots, r$ and $j = 1, \ldots, i$ and all the empty positions above are filled in with zero matrices. When a matrix A is in block triangular form, solving $Ax = b$ is reduced to solving smaller linear systems $A_{i,i}x_i = b_i$ after substitutions of partial solutions $\overline{x}_1, \ldots, \overline{x}_{i-1}$ of previously solved linear systems.

Methods have been proposed to transform a square matrix A into block triangular form with only permutations of rows and columns [9, Chap. 6]. These methods mainly consist of two steps: (1) apply permutations to A to result in non-zero diagonals if possible; and (2) apply symmetric permutations to result in a matrix in block triangular form. Steps (1) and (2) can be realized by Duff's algorithm in [8] and Tarjan's algorithm [26] respectively.

The algorithm due to Duff for transforming a matrix into non-zero diagonals will fail if the input matrix is so-called structurally singular. For example, when A contains a row full of zeros, it is impossible to transform A with only permutations of rows and columns to result in a matrix with non-zero diagonals. In the case of a structurally singular matrix, the algorithm due to Duff will return a matrix with the most non-zero diagonals.

Tarjan's algorithm is originally for finding the strongly connected components of directed graphs. When applied to a square matrix A, Tarjan's algorithm will find the symmetric permutations to transform A into block triangular form. Tarjan's algorithm is of low computational complexity $(O(\tau) + O(n)$ for a matrix of order n with τ entries) and easy to implement with stacks. Tarjan's algorithm assumes an input matrix with non-zero diagonals.

A trivial block triangular form of a matrix A is that there is only one block, and in this worst case there is no saving for solving $Ax = b$. When the matrix

A is dense, or equivalently the variables in $A\boldsymbol{x}$ are all heavily coupled, Tarjan's algorithm applied to A may result in such a trivial block triangular form. Therefore, the effectiveness of Tarjan's algorithm, which can be described by the number of blocks and sizes of the blocks, is dependent on the sparsity of the input matrix. Generally speaking, the sparser the matrix is, the more effective Tarjan's algorithm is.

3 Symbolic Detection of Steady States of Autonomous Differential Systems in Biology

3.1 Autonomous Differential Systems and Their Steady States

Consider the following n-dimensional autonomous differential system

$$\begin{cases} \dfrac{\mathrm{d}x_1}{\mathrm{d}t} = \dfrac{P_1(u_1,\ldots,u_m,x_1,\ldots,x_n)}{Q_1(u_1,\ldots,u_m,x_1,\ldots,x_n)}, \\[2ex] \qquad\vdots \\[2ex] \dfrac{\mathrm{d}x_n}{\mathrm{d}t} = \dfrac{P_n(u_1,\ldots,u_m,x_1,\ldots,x_n)}{Q_n(u_1,\ldots,u_m,x_1,\ldots,x_n)}, \end{cases} \qquad (2)$$

where $u_1,\ldots,u_m$ are parameters independent on t, $x_1,\ldots,x_n$ are variables dependent on t, and $P_1,\ldots,P_n,Q_1,\ldots,Q_n \in \mathbb{R}[u_1,\ldots,u_m,x_1,\ldots,x_n]$ are polynomials over the real field $\mathbb{R}$. We denote $\boldsymbol{u} = (u_1,\ldots,u_m)$ for simplicity.

Definition 3. For an arbitrary value $\overline{u} \in \mathbb{R}^m$ of the parameters $\boldsymbol{u}$, a point $\overline{x} = (\overline{x}_1,\ldots,\overline{x}_n) \in \mathbb{R}^n$ is said to be a *steady state* or *equilibrium* of system (2) if $P_1(\overline{u},\overline{x}) = \cdots = P_n(\overline{u},\overline{x}) = 0$ and $Q_1(\overline{u},\overline{x}) \neq 0, \cdots, Q_n(\overline{u},\overline{x}) \neq 0$.

Many biological systems like chemical reaction networks and reconstructed signal pathways can be modeled as (2), and the steady states and their stability of such systems are of our interest. For detection of steady states of the dynamic system (2), we need to first calculate all the real solutions $\varPhi \subset \mathbb{R}^n$ of $P_1(\boldsymbol{u},\boldsymbol{x}) = \cdots = P_n(\boldsymbol{u},\boldsymbol{x}) = 0$ with respect to $\boldsymbol{x}$ and then remove a solution $\overline{x} \in \varPhi$ if there exists some $\overline{u} \in \mathbb{R}^m$ and Q_j such that $Q(\overline{u},\overline{x}) = 0$. Next we focus on the first step by transforming the polynomial set $P_1,\ldots,P_n$ into block triangular form and then solving the polynomial sets blockwise. This transformation into block triangular form is achieved with Tarjan's algorithm applied to the adjacency matrix of $\mathcal{P}$ defined below, similar to the linear case.

For a polynomial $F \in \mathbb{K}[\boldsymbol{x}]$, the set of the variables which effectively appear in F is called the *(variable) support* of $\mathcal{F}$ and denoted by $\mathrm{supp}(F)$. Let $\mathcal{F} = [F_1,\ldots,F_n] \subset \mathbb{K}[\boldsymbol{x}]$ be a square polynomial set. Then the *adjacency matrix* M of $\mathcal{F}$ is an $n \times n$ matrix such that

$$\mathsf{M}_{i,j} = \begin{cases} 1, & \text{if } x_j \in \mathrm{supp}(F_i); \\ 0, & \text{otherwise.} \end{cases}$$

We call $\mathcal{F}$ a *sparse* polynomial system if its adjacency matrix is sparse. In particular, let $\mathsf{M}_{n \times n}$ be a matrix over $\mathbb{K}$. Then the *associated directed graph* $G(\mathsf{M}) = (V, E)$ of M is a directed graph such that $V = \{x_1, \ldots, x_n\}$ and $E = \{(x_j, x_i) : \mathsf{M}_{i,j} \neq 0\}$.

Tarjan's algorithm applied to the adjacency matrix M of a polynomial set $\mathcal{P}$ will return symmetric permutations of rows and columns to transform M. These permutations, considered in the settings of solving $\mathcal{P} = 0$, are merely reordering of the variables $x_1, \ldots, x_n$ and the polynomial equations $P_1 = 0, \ldots, P_n = 0$ and they will result in a new polynomial set $\mathcal{P}'$ equivalent to $\mathcal{P}$. In particular, the following observations make this strategy very suitable for autonomous differential systems arising from biology.

Let $\mathcal{P} = \{P_1, \ldots, P_n\} \subset \mathbb{K}[\boldsymbol{x}]$ as in (2). Then clearly $\mathcal{P}$ is a square polynomial set of n polynomials in n variables and thus permits a square adjacency matrix M of size $n \times n$. Furthermore, for polynomial systems arising from biology, the numbers of variables effectively appearing in them are large (easily tens of or even hundreds of variables), but the couplings of their variables in the polynomial systems from biology are quite loose in general, which results in a sparse adjacency matrix M. For examples, in the two biological dynamic systems we consider later in Sect. 4, the percentages of non-zero entries in the adjacency matrices are $29/12^2 \approx 20.1\%$ and $87/43^2 \approx 4.7\%$ respectively. The sparsity of M gives a larger chance for Tarjan's algorithm to transform $\mathcal{P}$ into smaller blocks, greatly reducing the complexity of solving $\mathcal{P} = 0$ symbolically.

When the adjacency matrix M of $\mathcal{P}$ has zero diagonals, Tarjan's algorithm applied to M still works as if M is of all non-zero diagonals. Suppose that $\mathsf{M}_{i,i} = 0$ for some i $(1 \leq i \leq n)$. Next we show that in this case $\{i\}$ will be one block in the result returned by Tarjan's algorithm applied to M. Suppose that $B = \{j_1, \ldots, j_s, i\}$ is a block returned by Tarjan's algorithm applied to a matrix M with $\mathsf{M}_{i,i} = 0$ for some i. This means that for each variable $x_k \in \mathrm{supp}(P_{j_1}) \cup \cdots \cup \mathrm{supp}(P_{j_s}) \cup \mathrm{supp}(P_i)$, either k is contained in the preceding blocks returned by Tarjan's algorithm or $k \in \{j_1, \ldots, j_s, i\}$. Since $\mathsf{M}_{i,i} = 0$, we have $x_i \notin \mathrm{supp}(P_i)$, and thus for each variable $x_k \in \mathrm{supp}(P_i)$, either k is contained in preceding blocks or $k \in \{j_1, \ldots, j_s\}$, and thus $B_1 = \{j_1, \ldots, j_s\}$ and $B_2 = \{i\}$ are two smaller blocks, which contradicts the fact that Tarjan's algorithm returns irreducible blocks.

In the case when $x_i \notin \mathrm{supp}(P_i)$ for some i $(1 \leq i \leq n)$, or namely $\mathsf{M}_{i,i} = 0$. Let $X = [\boldsymbol{x}_1, \ldots, \boldsymbol{x}_j = \{x_i\}, \ldots, \boldsymbol{x}_r]$ be a partition of the variables $\{x_1, \ldots, x_n\}$ returned by Tarjan's algorithm applied to M and $\Psi = [\mathcal{P}_1, \ldots, \mathcal{P}_j = \{P_i\}, \ldots, \mathcal{P}_r]$ be the corresponding polynomial partition of $\mathcal{P}$ in block triangular form with respect to X. Next we show how to refine the partitions X and $\mathcal{P}$ in this case. Since $x_i \notin \mathrm{supp}(P_i)$, we know that $\mathrm{supp}(P_i) \subset \cup_{l=1}^{j-1} \boldsymbol{x}_l$. Let $k = \max\{l \in \{1, \ldots, j-1\} : \exists x \in \mathrm{supp}(P_i) \text{ such that } x \in \boldsymbol{x}_l\}$. Then clearly we have $\mathrm{supp}(P_i) \subset \cup_{l=1}^{k} \boldsymbol{x}_l$. Let

$$X' = [\boldsymbol{x}_1, \ldots, \boldsymbol{x}_k \cup \{x_i\}, \ldots, \boldsymbol{x}_{j-1}, \boldsymbol{x}_{j+1}, \ldots, \boldsymbol{x}_r],$$
$$P' = [\mathcal{P}_1, \ldots, \mathcal{P}_k \cup \{P_i\}, \ldots, \mathcal{P}_{j-1}, \mathcal{P}_{j+1}, \ldots, \mathcal{P}_r].$$

Then it is easy to see that P' is in block triangular form with respect to X' and we finish the update for the integer i. To summarize, for each i $(1 \leq i \leq n)$ such that $x_i \notin \mathrm{supp}(P_i)$, we will update the variable and polynomial partitions X and $\mathcal{P}$ accordingly before we start the process of solving the resultant polynomial systems blockwise.

3.2   A Refined Algorithm for Solving Square Polynomial Systems by Transformation into Block Triangular Form

Based on the discussions above, we formulate the procedure for solving square polynomial systems by using Tarjan's algorithm for transforming the polynomial set into block triangular form as Algorithm 1 below. In this algorithm, the subroutine $\mathsf{Tarjan}(\cdot)$ takes a square matrix and returns an ordered partition $[x_1, \ldots, x_r]$ for some r of $\{x_1, \ldots, x_n\}$ such that the input matrix is in block triangular form with respect to this partition, and the subroutine $\mathsf{Solve}(\mathcal{F}, x)$ solves the polynomial system $\mathcal{F} = 0$ with respect to the variables x symbolically (for example by using triangular decomposition or computation of Gröbner bases) and returns all the solutions as a set. The operator $\mathrm{index}(x_i)$ returns the index i of the input variable x_i, and $[A]\mathrm{cat}[B]$ equals $[A, B]$.

With Algorithm 1, finding all the steady states of autonomous differential biological systems modeled as (2) can be achieved more effectively with $\mathsf{BlockTriangular}(\cdot)$, for a biological system usually furnishes a sparse adjacency matrix when n is large.

4   Two Illustrative Examples

In this section we illustrate the effectiveness of finding steady states of autonomous differential biological systems by transformation into block triangular forms with two biological models.

4.1   Synthesis of One Enzyme in Bacterial Cells: 12 Variables

Consider the following autonomous differential equation system in [3] which describes the synthesis of one enzyme in bacterial cells.

$$dx_1/dt = p_1 x_3 - (p_2 + p_3)x_1, \qquad dx_2/dt = p_{15}x_3 - p_4 x_2,$$
$$dx_3/dt = (p_2 + p_3)x_1 - (p_1 + p_{15})x_3 + p_4 x_2, \qquad dx_4/dt = p_{12}x_7 x_6 - p_7 x_4$$
$$dx_5/dt = p_6 x_7 - p_8 x_5, \qquad dx_6/dt = p_3 x_1 - p_5 x_6 - p_{12}x_6 x_7,$$
$$dx_7/dt = -p_{12}x_6 x_7 + p_7 x_4 - p_6 x_7 + p_8 x_5, \qquad dx_8/dt = p_{14}x_{12} - p_{13}x_8 x_9,$$
$$dx_9/dt = p_8 x_5 - p_9 x_9 - p_{13}x_8 x_9, \qquad dx_{10}/dt = p_{11}x_{11},$$
$$dx_{11}/dt = p_{10}x_{12} - p_{11}x_{11}, \qquad dx_{12}/dt = p_{13}x_8 x_9 + p_{11}x_{11} - (p_{10}+p_{14})x_{12}.$$

In these differential equations t is the temporal variable, $x_1, \ldots, x_{12}$ are the variables dependent on t, and $p_1, \ldots, p_{14}$ are constants. Denote the polynomials

---

Algorithm 1: Algorithm for solving square polynomial systems by transformation into block triangular form $\Phi :=$ BlockTriangular($\mathcal{F}$)

Input: $\mathcal{F} = [F_1, \ldots, F_n]$, a square polynomial set in $\mathbb{K}[\boldsymbol{x}]$
Output: Φ, all the solutions of $\mathcal{F} = 0$ in $\overline{\mathbb{K}}^n$

1 M := the adjacency matrix of $\mathcal{F}$;
2 $Z := \{i \in \{1, \ldots, n\} : \mathsf{M}_{i,i} = 0\}$;
3 $X :=$ Tarjan(M);   *Compute the ordered variable partition*
4 $\mathcal{P} := [\,]$;   *Construct the corresponding polynomial partition*
5 **for** $\tilde{x} \in X$ **do**
6     $\mathcal{I} := \{\mathrm{index}(x_i) : x_i \in \tilde{x}\}$;
7     $\mathcal{P} := \mathcal{P}$ cat $[\{F_i \in \mathcal{F} : i \in \mathcal{I}\}]$;

8 **for** $i \in Z$ **do**
9     $\mathcal{B} := \emptyset$;
10   **for** $x \in \mathrm{supp}(F_i)$ **do**
11   Find j s.t. $x \in X[j]$; $\mathcal{B} := \mathcal{B} \cup \{j\}$;

12    $k := \max(\mathcal{B})$;
13   Remove $\{F_i\}$ from $\mathcal{P}$ and $\{x_i\}$ from X;   *Update the partitions X and $\mathcal{P}$*
14    $\boldsymbol{x}_k := \boldsymbol{x}_k \cup \{x_i\}$, $\mathcal{P}_k := \mathcal{P}_k \cup \{F_i\}$;

15 $\Phi := \{[\,]\}$;   *Solving $\mathcal{F} = 0$ blockwise*
16 **for** $i = 1, \ldots, \#\mathcal{P}$ **do**
17    $\Phi_t := \{[\,]\}$;
18   **for** $\overline{x} \in \Phi$ **do**
19       $\Psi :=$ Solve($\mathcal{P}[i](\overline{x}, X[i]), X[i]$);
20   **if** $\Psi = \emptyset$ **then**
21   **break**;
22   **else**
23          $\Phi_t := \Phi_t \cup \{\overline{x} \text{ cat } \overline{x}' : \overline{x}' \in \Psi\}$;

24    $\Phi := \Phi_t$;

25 **return** Φ;

---

in the right-hand side of the equations above by $P_1, \ldots, P_{12}$ respectively. Then the adjacency matrix M of $\mathcal{P} = \{P_1, \ldots, P_{12}\}$ is

$$
M = \begin{pmatrix}
1 & 0 & 1 & 0 & 0 & 1 & 0 & 0 & 0 & 0 & 0 & 0 \\
0 & 1 & 1 & 0 & 0 & 0 & 0 & 0 & 0 & 0 & 0 & 0 \\
1 & 1 & 1 & 0 & 0 & 0 & 0 & 0 & 0 & 0 & 0 & 0 \\
0 & 0 & 0 & 1 & 0 & 0 & 1 & 0 & 0 & 0 & 0 & 0 \\
0 & 0 & 0 & 0 & 1 & 0 & 1 & 1 & 1 & 0 & 0 & 0 \\
0 & 0 & 0 & 1 & 0 & 1 & 0 & 1 & 0 & 0 & 0 & 0 \\
0 & 0 & 0 & 1 & 1 & 1 & 1 & 0 & 0 & 0 & 0 & 0 \\
0 & 0 & 0 & 0 & 0 & 0 & 1 & 1 & 0 & 0 & 0 & 0 \\
0 & 0 & 0 & 0 & 0 & 0 & 1 & 1 & 0 & 0 & 0 & 0 \\
0 & 0 & 0 & 0 & 0 & 0 & 0 & 0 & 0 & 0 & 0 & 0 \\
0 & 0 & 0 & 0 & 0 & 0 & 0 & 0 & 0 & 1 & 1 & 1 \\
0 & 0 & 0 & 0 & 0 & 0 & 0 & 0 & 0 & 0 & 1 & 1
\end{pmatrix}.
$$

One may find that the 10-th row of M is a zero one, and thus x_{10} does not appear in any polynomial in $\mathcal{P}$. The associated directed graph of M is shown in Fig. 1 below. A clear structure in the graph of three blocks can be found visually.

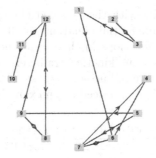

Fig. 1. Associated directed graph: 12 variables

Tarjan's algorithm applied to M returns the ordered partition

$$X = [\{1,3,2\}, \{4,6,7,5\}, \{8,9,12,11\}, \{\underline{10}\}],$$

where $\{\underline{10}\}$ means that the variable x_{10} does not appear in P_{10}. Since P_{10} only involves one variable x_{11}, the variable and polynomial partitions are updated as

$$[\{1,3,2\}, \{4,6,7,5\}, \{8,9,12,11,10\}],$$
$$[\{F_1, F_3, F_2\}, \{F_4, F_6, F_7, F_5\}, \{F_8, F_9, F_{12}, F_{11}, F_{10}\}]$$

respectively.

Solving the first polynomial block $F_1 = F_2 = F_3 = 0$ with respect to the first variable block $\{x_1, x_2, x_3\}$ results in one partial solution

$$x_1 = x_1, \quad x_2 = \frac{p_{15}(p_2 + p_3)x_1}{p_1 p_4}, \quad x_3 = \frac{(p_2 + p_3)x_1}{p_1}.$$

After successive substitutions of the partial solution to and then solving the second and third polynomial blocks, this partial solution is extended to the solution of $\mathcal{F} = 0$:

$$x_1 = x_1, \quad x_2 = \frac{p_{15}(p_2 + p_3)x_1}{p_1 p_4}, \quad x_3 = \frac{(p_2 + p_3)x_1}{p_1}, \quad x_4 = \frac{p_3 x_1 - p_5 x_6}{p_7},$$
$$x_5 = \frac{p_6(p_3 x_1 - p_5 x_6)}{(p_{12} x_6 p_8)}, \quad x_6 = x_6, \quad x_7 = \frac{p_3 x_1 - p_5 x_6}{p_{12} x_6}, \quad x_9 = \frac{p_6(p_3 x_1 - p_5 x_6)}{p_9 p_{12} x_6},$$
$$x_8 = x_{11} = x_{12} = 0,$$

where x_1 and x_6 are free variables. Any combination of specific values for x_1 and x_6 (with $x_6 \neq 0$) leads to a specific solution of $\mathcal{F} = 0$.

4.2 Signaling Network for Leishmaniasis with Positive Feedback Loop: 43 Variables

A medium-size example we consider is a reconstructed mathematical signaling network for leishmaniasis with positive feedback loop in [21], also deposited in the BioModels Database [18] with the assigned identifier MODEL1308080000. This model consists of 43 differential equations in 43 variables for the interacting species to describe the reactions in the reconstructed network, and all the data, including the values for the kinetic parameters, the equations, and initial conditions can be found in the online supplementary data of the paper. Below one equation in the model is reproduced as an example:

$$\frac{d[MKK4/7]V_c}{dt} = V_c \frac{0.3[TAB2\_TAK1\_TAB1]}{0.01 + [TAB2\_TAK1\_TAB1]V_c} + V_c \frac{0.4[Ras]}{1.5 + [Ras]V_c}$$
$$+ V_c \frac{0.6[MEKK1]}{0.2 + [MEKK1]V_c} - V_c \frac{0.98[MKK4/7]}{0.15 + [MKK4/7]V_c},$$

where $[MKK4/7]$, $[TAB2\_TAK1\_TAB1]$, $[Ras]$, and $[MEKK1]$ are the variables for species in [21]. We rename the variables in the left hands of the differential equations to $x_1, \ldots, x_{43}$ as shown in the following table (Table 1).

Table 1. Variable correspondences

Species	TNFc	TNFR1	TRADD_TRAF2_RIP	IkB	NIK	IkK_NFkB	NFkBc	MEKK1
Variables	x_1	x_2	x_3	x_4	x_5	x_6	x_7	x_8
Species	MKK4/7	JNKc	ASK	p38c	LPG	CD14-TLR	MyD88	IRAK1/4
Variables	x_9	x_{10}	x_{11}	x_{12}	x_{13}	x_{14}	x_{15}	x_{16}
Species	TRAF6	TAB2_TAK1_TAB1	MKK1/2	ERK1/2c	MKK3/6	EGF	EGFR	PLC gamma
Variables	x_{17}	x_{18}	x_{19}	x_{20}	x_{21}	x_{22}	x_{23}	x_{24}
Species	PIP2	DAG	PKC	PI3K	Aktc	Shc/Grb2/Sos1	Ras	Raf
Variables	x_{25}	x_{26}	x_{27}	x_{28}	x_{29}	x_{30}	x_{31}	x_{32}
Species	JAK	STAT1/3c	NFkBn	JNKn	cjun	p38n	cfos	ERK1/2n
Variables	x_{33}	x_{34}	x_{35}	x_{36}	x_{37}	x_{38}	x_{39}	x_{40}
Species	Aktn	STAT1/3n	TNFn					
Variables	x_{41}	x_{42}	x_{43}					

The adjacency matrix M of the polynomial system is a 43×43 matrix with 87 non-zero entries. The associated directed graph of this adjacency matrix is shown in Fig. 2 below. Note that some of the adjacent vertexes in the right circle are also connected but the edges are not clearly shown.

The application of Tarjan's algorithm to M returns the following partition

[{13},{14},{15},{16},{17},{18},{22},{23},{24},{25},{26},{27},{1,43,35,7,6,4,5,3,2},{8},{30},{31}, {9},{10},{11},{21},{12},{32},{19},{20},{28},{29},{33},{34},{36},{37},{38},{39},{40},{41},{42}],

and after the update the variable partition becomes

[{13},{14},{15},{16},{17},{18},{22},{23},{24},{25},{26},{27},{1,43,35,7,6,4,5,3,2},{8},{30}, {31},{9},{10},{11},{21},{12},{32},{19},{20},{28},{29,41},{33},{34,42},{36,37},{38,39}, {40}],

and the polynomial partition also gets updated accordingly.

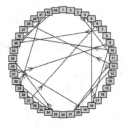

Fig. 2. Associated directed graph: 43 variables

This partition indicates that following the specific ordering as shown in the above partition, the essential difficulty of solving the original polynomial system of 43 variables is reduced to solving one polynomial system of 9 variables (corresponding to the block $\{1, 43, 35, 7, 6, 4, 5, 3, 2\}$) and some univariate polynomials after substitutions of previously computed partial solutions. As discussed in the introduction, this greatly reduces the cost for solving this polynomial set.

5   Concluding Remarks

Autonomous differential systems are typical mathematical models of many biological systems like chemical reaction networks. In the study of steady states and their stability of such systems, solving polynomial systems symbolically is one important step but its efficiency suffers when the numbers of variables in the systems are large.

In this paper we propose a method for transforming a square polynomial set into block triangular form by using Tarjan's algorithm applied to the adjacency matrix of the polynomial set. This method can be viewed as a preprocessing step for solving polynomial systems with little cost compared with the actual solving, and it is effective when the polynomial system admits a sparse adjacency matrix.

Biological systems are perfect candidates for the applications of this method, for the variable numbers in such systems are usually large and the systems themselves are sparse due to the loose coupling of the variables. Two biological systems of 12 and 43 variables respectively are studied to illustrate the effectiveness of the proposed method.

Acknowledgments. The author would like to thank Yufei Gao and Yishan Cui for their help in the investigation on Tarjan's algorithm and the biological database and the anonymous reviewers for their helpful comments which lead to improvement on this manuscript and potential enrichment in its extended version.

References

1. Allen, L.J.: Some discrete-time SI, SIR, and SIS epidemic models. Math. Biosci. **124**(1), 83–105 (1994)
2. Aubry, P., Lazard, D., Moreno Maza, M.: On the theories of triangular sets. J. Symbolic Comput. **28**(1–2), 105–124 (1999)
3. Bock, H.G.: Numerical treatment of inverse problems in chemical reaction kinetics. In: Ebert, K.H., Deuflhard, P., Jäger, W. (eds.) Modelling of Chemical Reaction Systems. Springer Series in Chemical Physics, vol. 18, pp. 102–125. Springer, Heidelberg (1981). https://doi.org/10.1007/978-3-642-68220-9_8
4. Buchberger, B.: Ein Algorithmus zum Auffinden der Basiselemente des Restklassenrings nach einem nulldimensionalen Polynomideal. Ph.D. thesis, Universität Innsbruck, Austria (1965)
5. Cifuentes, D., Parrilo, P.A.: Exploiting chordal structure in polynomial ideals: A Gröbner bases approach. SIAM J. Discrete Math. **30**(3), 1534–1570 (2016)
6. Collins, G.E.: Quantifier elimination for real closed fields by cylindrical algebraic decompostion. In: Brakhage, H. (ed.) Automata Theory and Formal Languages 2nd GI Conference. LNCS, vol. 33, pp. 134–183. Springer, Heidelberg (1975). https://doi.org/10.1007/3-540-07407-4_17
7. Collins, G.E., Hong, H.: Partial cylindrical algebraic decomposition for quantifier elimination. J. Symbolic Comput. **12**(3), 299–328 (1991)
8. Duff, I.S.: On algorithms for obtaining a maximum transversal. ACM Trans. Math. Softw. **7**(3), 315–330 (1981)
9. Duff, I.S., Erisman, A.M., Reid, J.K.: Direct Methods for Sparse Matrices. Oxford University Press, New York (1986)
10. El Kahoui, M., Weber, A.: Deciding Hopf bifurcations by quantifier elimination in a software-component architecture. J. Symbolic Comput. **30**(2), 161–179 (2000)
11. Faugère, J.C.: A new efficient algorithm for computing Gröbner bases (F_4). J. Pure Appl. Algebra **139**(1–3), 61–88 (1999)
12. Faugère, J.C., Mou, C.: Sparse FGLM algorithms. J. Symbolic Comput. **80**(3), 538–569 (2017)
13. Faugère, J.C., Rahmany, S.: Solving systems of polynomial equations with symmetries using SAGBI-Gröbner bases. In: May, J.P. (ed.) Proceedings of ISSAC 2009, pp. 151–158. ACM (2009)
14. Faugère, J.C., Spaenlehauer, P.J., Svartz, J.: Sparse Gröbner bases: The unmixed case. In: Nabeshima, K., Nagasaka, K. (eds.) Proceedings of ISSAC 2014, pp. 178–185. ACM (2014)
15. Ferrell, J.E., Tsai, T.Y.C., Yang, Q.: Modeling the cell cycle: why do certain circuits oscillate? Cell **144**(6), 874–885 (2011)
16. Gatermann, K., Huber, B.: A family of sparse polynomial systems arising in chemical reaction systems. J. Symbolic Comput. **33**(3), 275–305 (2002)
17. Laubenbacher, R., Sturmfels, B.: Computer algebra in systems biology. Am. Math. Monthly **116**(10), 882–891 (2009)
18. Li, C., Donizelli, M., Rodriguez, N., Dharuri, H., et al.: Biomodels database: an enhanced, curated and annotated resource for published quantitative kinetic models. BMC Syst. Biol. **4**(1), 92 (2010)
19. Li, X., Mou, C., Niu, W., Wang, D.: Stability analysis for discrete biological models using algebraic methods. Math. Comput. Sci. **5**(3), 247–262 (2011)
20. Mayr, E., Meyer, A.: The complexity of the word problems for commutative semigroups and polynomial ideals. Adv. Math. **46**(3), 305–329 (1982)

21. Mol, M., Patole, M.S., Singh, S.: Immune signal transduction in leishmaniasis from natural to artificial systems: role of feedback loop insertion. Biochim. Biophys. Acta Gen. Subj. **1840**(1), 71–79 (2014)

22. Mou, C., Bai, Y.: On the chordality of polynomial sets in triangular decomposition in top-down style (2018). arXiv:1802.01752

23. Niu, W., Wang, D.: Algebraic analysis of bifurcation and limit cycles for biological systems. In: Horimoto, K., Regensburger, G., Rosenkranz, M., Yoshida, H. (eds.) AB 2008. LNCS, vol. 5147, pp. 156–171. Springer, Heidelberg (2008). https://doi.org/10.1007/978-3-540-85101-1_12

24. Niu, W., Wang, D.: Algebraic approaches to stability analysis of biological systems. Math. Comput. Sci. **1**(3), 507–539 (2008)

25. Sturm, T., Weber, A., Abdel-Rahman, E.O., El Kahoui, M.: Investigating algebraic and logical algorithms to solve Hopf bifurcation problems in algebraic biology. Math. Comput. Sci. **2**(3), 493–515 (2009)

26. Tarjan, R.: Depth-first search and linear graph algorithms. SIAM J. Comput. **1**(2), 146–160 (1972)

27. Wang, D.: Elimination Methods. Springer, Heidelberg (2001). https://doi.org/10.1007/978-3-7091-6202-6

28. Wang, D., Xia, B.: Stability analysis of biological systems with real solution classification. In: Kauers, M. (ed.) Proceedings of ISSAC 2005, pp. 354–361. ACM Press (2005)

29. Yang, L., Xia, B.: Real solution classifications of parametric semi-algebraic systems. In: Proceedings of A3L 2005, pp. 281–289. Herstellung und Verlag, Norderstedt (2005)

Consensus Decoding of Recurrent Neural Network Basecallers

Jordi Silvestre-Ryan and Ian Holmes[✉]

Department of Bioengineering, University of California, Berkeley, USA
{jordisr,ihh}@berkeley.edu

Abstract. There is an extensive literature using probabilistic models, such as hidden Markov models, for the analysis of biological sequences. These models have a clear theoretical basis, and many heuristics have been developed to reduce the time and memory requirements of the dynamic programming algorithms used for their inference. Nevertheless, mirroring the shift in natural language processing, bioinformatics is increasingly seeing higher accuracy predictions made by recurrent neural networks (RNN). This shift is exemplified by basecalling on the Oxford Nanopore Technologies' sequencing platform, in which a continuous time series of current measurements is mapped to a string of nucleotides. Current basecallers have applied connectionist temporal classification (CTC), a method originally developed for speech recognition, and focused on the task of decoding RNN output from a single read. We wish to extend this method for the more general task of consensus basecalling from multiple reads, and in doing so, exploit the gains in both accelerated algorithms for sequence analysis and recurrent neural networks, areas that have advanced in parallel over the past decade. To this end, we develop a dynamic programming algorithm for consensus decoding from a pair of RNNs, and show that it can be readily optimized with the use of an alignment envelope. We express this decoding in the notation of finite state automata, and show that pair RNN decoding can be compactly represented using automata operations. We additionally introduce a set of Markov chain Monte Carlo moves for consensus basecalling multiple reads.

Keywords: Nanopore sequencing · Deep learning
Dynamic programming · Alignment envelope · Finite state automata

1 Introduction

In nanopore sequencing, such as that done on the MinION platform produced by Oxford Nanopore Technologies (ONT), a single strand of DNA is passed through a protein nanopore on a synthetic membrane. A voltage is applied across the membrane, such that as the DNA goes through the pore, it alters the electrical current in a sequence-dependent way. The basecaller then uses a learned model

© Springer International Publishing AG, part of Springer Nature 2018
J. Jansson et al. (Eds.): AlCoB 2018, LNBI 10849, pp. 128–139, 2018.
https://doi.org/10.1007/978-3-319-91938-6_11

to find the most probable nucleotide sequence given the input of current measurements. Originally this was done with hidden Markov models (e.g. Nanocall [2]), though more recent basecallers produced both by ONT and the academic community have moved to recurrent neural networks (RNNs).

First applied to the speech recognition, which similarly seeks to find a mapping between a continuous input and a discrete set of labels, connectionist temporal classification (CTC) uses a loss function and associated decoding algorithm that allow for gaps and variable spacing of signal [4]. The ONT basecalling tool Scrappie is an early example of CTC applied to bioinformatic RNNs. CTC is implemented in the TensorFlow library, and has been used by other nanopore basecallers, e.g. Chiron [10].

In Sect. 2, we present an algorithm for consensus CTC decoding of paired RNNs. The algorithm, which is not tied to any particular RNN architecture and may be used quite generally, makes use of the "alignment envelope" technique from bioinformatics sequence analysis. In this technique, a set of plausible candidate alignments is quickly estimated for a pair of sequences, then applied as a filter mask to more complex calculations [5,6]. This is particularly relevant for the $1D^2$ sequencing protocol, in which a DNA strand and its complement are both passed through a nanopore successively, thus yielding two current traces over the same sequence.

In Sect. 3, we present an overview of an alternative consensus approach using Markov chain Monte Carlo. Besides decoding $1D^2$ traces, this algorithm can be easily extended to more than two sequences, making it useful for consensus basecalling in general.

Finally, in Sect. 4, we show how our dynamic programming algorithms may be represented using an algebra of intersecting finite-state automata[1]. This algebraic framework is not necessary to follow or implement our algorithms, but may help in the systematic derivation and verification of such algorithms.

2   Pair Decoding of RNNs

2.1   Recurrent Neural Network

Following a notation similar to [4], consider a recurrent neural network with input sequence $\mathbf{x}$ having length T, and an output matrix of probabilities $\mathbf{y}$ having dimension $T \times (|L| + 1)$, where L is a label alphabet.

The matrix $\mathbf{y}$ represents a gapped position-specific weight matrix profile, and can be thought of as a simple automaton (Fig. 1). Element y_a^t is the probability that the t'th position has output a, whether this is a label character ($a \in L$) or a gap ($a = \epsilon$).

---

[1] This builds on the interpretation of Scrappie, and similar CTC-decoding basecallers, as "transducer" neural networks (Tim Massingham, Oxford Nanopore Technologies, *pers. comm.*).

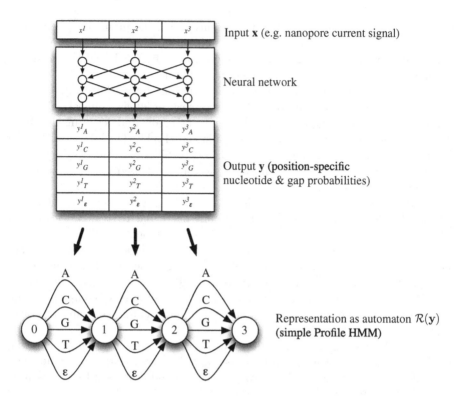

Fig. 1. Representation of neural network outputs as a hidden Markov model. The transition weights for the state machine $\mathcal{R}(\mathbf{y})$ (bottom) are given by the output probability matrix $\mathbf{y}$ of the RNN. The neural network may have any architecture, as long as the output matrix has the form shown here.

2.2   Overview of CTC Decoding

The idea of Connectionist Temporal Classification decoding (CTC) is to find the most likely label sequence ℓ, marginalizing away the gaps. In order to find this sequence we need to be able to (efficiently) compute two probabilities:

- the probability $P(\ell|\mathbf{x})$ that ℓ is the correct label sequence;
- the probability $\sigma(\kappa|\mathbf{x})$ that κ is a prefix of the correct label sequence.

These probabilities, which both sum over (potentially) many ways that $\mathbf{y}$ can encode ℓ, are computed by dynamic programming using the Forward algorithm. The derivatives of $P(\ell|\mathbf{x})$ with respect to the RNN parameters, which are required for training, can be computed using the Forward-Backward algorithm.

Assuming that we can calculate these probabilities, CTC finds the most likely label sequence by a prefix search. This can be accomplished as follows:

- Maintain a priority queue of prefix sequences κ, sorted by their prefix probability $\sigma(\kappa|\mathbf{x})$.
- At each step, the top candidate is extended by one label symbol, trying each of the symbols from the label alphabet; the $|L|$ different extended prefix sequences that are so created are added onto the priority queue.
- For each sequence κ that is visited as a prefix, the probability $P(\ell = \kappa|\mathbf{x})$ of it also being the exact correct label is calculated, and the most probable such label $\hat{\ell}$ is recorded.
- The search terminates when the probability $\sigma(\kappa|\mathbf{x})$ of the best unextended prefix κ is less than the label probability $P(\ell = \hat{\ell}|\mathbf{x})$ of the best label $\hat{\ell}$.

In practice, some additional constraints are required to prevent the prefix search space from exploding [4].

2.3 Dynamic Programming Calculations

Let $\pi \in (L')^T$ be a *path*, comprising T symbols from the extended label alphabet $L' = L \cup \{\epsilon\}$. Define the path probability

$$P(\pi|\mathbf{x}) = \prod_{t=1}^{T} y_{\pi_t}^t$$

Let $\mathbb{B}(\pi)$ be the function that removes all ϵ's from π, converting a path to a label sequence. In the original CTC description this function also removes repeated label characters, though here we consider a simplified version. Let $\ell_{i:j}$ denote the subsequence of ℓ from position i to position j (inclusive), or the empty sequence ϵ if $i > j$.

Let t denote an index into $\mathbf{y}$ and let s denote an index into ℓ.

The Forward-Backward dynamic programming recursions are, for $1 \leq t \leq T$ and $0 \leq s \leq |\ell|$

$$\alpha_t^\epsilon(\ell_{1:s}) = y_\epsilon^t \alpha_{t-1}(\ell_{1:s})$$
$$\alpha_t^*(\ell_{1:s}) = y_{\ell_s}^t \alpha_{t-1}(\ell_{1:s-1})$$
$$\beta_t^\epsilon(\ell_{s:|\ell|}) = y_\epsilon^t \beta_{t+1}(\ell_{s:|\ell|})$$
$$\beta_t^*(\ell_{s:|\ell|}) = y_{\ell_s}^t \beta_{t+1}(\ell_{s+1:|\ell|})$$

where

$$\alpha_t(\ell_{1:s}) = \alpha_t^\epsilon(\ell_{1:s}) + \alpha_t^*(\ell_{1:s})$$
$$\beta_t(\ell_{s:|\ell|}) = \beta_t^\epsilon(\ell_{s:|\ell|}) + \beta_t^*(\ell_{s:|\ell|})$$

with appropriate boundary conditions $(\alpha_0(\epsilon) = \beta_{T+1}(\epsilon) = 1$, and $\alpha_0(\kappa) = \beta_{T+1}(\kappa) = 0$ for all other κ). These compute the following probabilities

$$\alpha_t^\epsilon(\ell_{1:s}) = \sum_{\pi \in \mathbb{P}_t^\epsilon(\ell_{1:s})} P(\pi|\mathbf{x})$$

$$\alpha_t^*(\ell_{1:s}) = \sum_{\pi \in \mathbb{P}_t^*(\ell_{1:s})} P(\pi|\mathbf{x})$$

$$\beta_t^\epsilon(\ell_{s:|\ell|}) = \sum_{\pi \in \mathbb{S}_{T-t+1}^\epsilon(\ell_{s:|\ell|})} P(\pi|\mathbf{x})$$

$$\beta_t^*(\ell_{s:|\ell|}) = \sum_{\pi \in \mathbb{S}_{T-t+1}^*(\ell_{s:|\ell|})} P(\pi|\mathbf{x})$$

where the Forward probabilities (α) are marginals over sets of *prefix paths* ($\mathbb{P}$) and the Backward probabilities (β) are marginals over sets of *suffix paths* ($\mathbb{S}$), with the path sets further partitioned by whether the paths end in a gap character ($\alpha^\epsilon, \beta^\epsilon; \mathbb{P}^\epsilon, \mathbb{S}^\epsilon$) or a label symbol ($\alpha^*, \beta^*; \mathbb{P}^*, \mathbb{S}^*$)

$$\mathbb{P}_t(\kappa) = \{\pi : \pi \in (L')^t, \mathbb{B}(\pi) = \kappa\}$$
$$\mathbb{P}_t^\epsilon(\kappa) = \{\pi : \pi \in (L')^t, \mathbb{B}(\pi) = \kappa, \pi_t = \epsilon\}$$
$$\mathbb{P}_t^*(\kappa) = \{\pi : \pi \in (L')^t, \mathbb{B}(\pi) = \kappa, \pi_t \neq \epsilon\}$$
$$\mathbb{S}_t^\epsilon(\kappa) = \{\pi : \pi \in (L')^t, \mathbb{B}(\pi) = \kappa, \pi_1 = \epsilon\}$$
$$\mathbb{S}_t^*(\kappa) = \{\pi : \pi \in (L')^t, \mathbb{B}(\pi) = \kappa, \pi_1 \neq \epsilon\}$$

The marginal probability of a given label sequence is

$$P(\ell|\mathbf{x}) = \sum_{\pi \in \mathbb{P}_T(\ell)} P(\pi|\mathbf{x})$$
$$= \alpha_T(\ell)$$
$$= \beta_1(\ell) \tag{1}$$

The derivatives of this probability (required for training) are

$$\frac{\partial}{\partial\theta} P(\ell|\mathbf{x}) = \sum_{s=0}^{|\ell|}\sum_{t=1}^{T} \alpha_{t-1}(\ell_{1:s}) \cdot \frac{\partial y_\epsilon^t}{\partial\theta} \cdot \beta_{t+1}(\ell_{s+1:|\ell|})$$

$$+ \sum_{s=1}^{|\ell|}\sum_{t=1}^{T} \alpha_{t-1}(\ell_{1:s-1}) \cdot \frac{\partial y_{\ell_s}^t}{\partial\theta} \cdot \beta_{t+1}(\ell_{s+1:|\ell|})$$

The marginal probability that the label sequence has prefix κ can be calculated by conditioning on the output position associated with the last symbol of κ and then summing this out

$$\sigma(\kappa|\mathbf{x}) = \sum_{t=1}^{T} \alpha_t^*(\kappa)$$

Efficient computation of the Forward probabilities during the prefix search relies on the fact that, if ℓ is a prefix of ℓ', then many of the α used in computing $P(\ell|\mathbf{x})$ can be reused when computing $P(\ell'|\mathbf{x})$.

2.4   Consensus of Two RNNs

Consider the case where we have two RNNs with respective input sequences $\mathbf{u}$, $\mathbf{v}$ (lengths U, V) and output probability matrices $\mathbf{y}$, $\mathbf{z}$ (sizes $U \times (|L| + 1)$ and $V \times (|L| + 1)$).

We can re-use the prefix search strategy, now using the following probabilities:

- the probability $P(\ell = \ell'|\mathbf{u}, \mathbf{v})$ that ℓ' is the correct label sequence;
- the probability $\sigma(\kappa|\mathbf{u}, \mathbf{v})$ that κ is a prefix of the correct label sequence.

We want to find the best labeling **given that both RNNs agree**, so we define the above probabilities for paired RNNs as follows

$$P(\ell = \ell'|\mathbf{u}, \mathbf{v}) = \frac{1}{Z} P(\ell = \ell'|\mathbf{u}) P(\ell = \ell'|\mathbf{v}) \tag{2}$$

$$\sigma(\kappa|\mathbf{u}, \mathbf{v}) = \sum_{\ell' \in L^*} P(\ell = \kappa \oplus \ell'|\mathbf{u}, \mathbf{v})$$

where Z is the probability that the RNNs agree (this is a normalization constant that does not depend on ℓ), and $\oplus$ denotes sequence concatenation.

As before, we introduce variables $\pi \in (L')^U$, $\phi \in (L')^V$ representing paths through the two models, with path probabilities

$$P(\pi|\mathbf{u}) = \prod_{u=1}^{U} y_{\pi_u}^u$$

$$P(\phi|\mathbf{v}) = \prod_{v=1}^{V} z_{\pi_v}^v$$

Probability that the Two RNNs Agree. The probability that the two RNNs agree on the same label sequence is $Z = \gamma_{1,1}$ where $\gamma_{u,v}$ is the probability that $\mathbf{y}_{u:U}$ agrees with $\mathbf{z}_{v:V}$

$$\gamma_{u,v}^\epsilon = y_\epsilon^u \gamma_{u+1,v}$$

$$\gamma_{u,v}^{*\epsilon} = z_\epsilon^v \gamma_{u,v+1}^*$$

$$\gamma_{u,v}^{**} = \sum_{a \in L} y_a^u z_a^v \gamma_{u+1,v+1}$$

$$\gamma_{u,v}^* = \gamma_{u,v}^{*\epsilon} + \gamma_{u,v}^{**}$$

$$\gamma_{u,v} = \gamma_{u,v}^\epsilon + \gamma_{u,v}^*$$

with boundary condition $\gamma_{U+1,V+1}^{**} = 1$ (and other "out-of-bounds" likelihoods defined to be zero). These recursions are illustrated in Fig. 2.

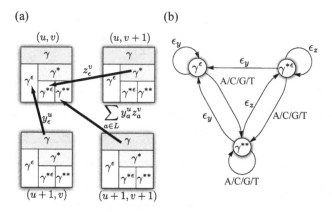

Fig. 2. (a) The recursions for calculating the γ dynamic programming matrix. The terms above each arrow are multiplied by the cell from which they originate to give the left hand side of each recursion. The shaded recursive variables at each time step are obtained by summing the cells directly under them. (b) Simple automaton illustrating the relationship between recursive variables. To avoid overcounting gapped paths, an ordering is imposed such that those from z appear first (shown by lack of an edge from γ^ϵ to $\gamma^{*\epsilon}$).

The interpretation of these probabilities is

$$\gamma_{u,v}^\epsilon = \sum_{(\pi,\phi)\in\mathbb{C}^\epsilon(u,v)} P(\pi|\mathbf{u})P(\phi|\mathbf{v})$$

$$\gamma_{u,v}^{*\epsilon} = \sum_{(\pi,\phi)\in\mathbb{C}^{*\epsilon}(u,v)} P(\pi|\mathbf{u})P(\phi|\mathbf{v})$$

$$\gamma_{u,v}^{**} = \sum_{(\pi,\phi)\in\mathbb{C}^{**}(u,v)} P(\pi|\mathbf{u})P(\phi|\mathbf{v})$$

where

$$\mathbb{C}^\epsilon(u,v) = \{(\pi,\phi) : \pi \in (L')^{U-u+1}, \phi \in (L')^{V-v+1}, \mathbb{B}(\pi) = \mathbb{B}(\phi), \pi_1 = \epsilon\}$$
$$\mathbb{C}^{*\epsilon}(u,v) = \{(\pi,\phi) : \pi \in (L')^{U-u+1}, \phi \in (L')^{V-v+1}, \mathbb{B}(\pi) = \mathbb{B}(\phi), \pi_1 \neq \epsilon, \phi_1 = \epsilon\}$$
$$\mathbb{C}^{**}(u,v) = \{(\pi,\phi) : \pi \in (L')^{U-u+1}, \phi \in (L')^{V-v+1}, \mathbb{B}(\pi) = \mathbb{B}(\phi), \pi_1 \neq \epsilon, \phi_1 \neq \epsilon\}$$

Analogously to $\mathbb{P}$ (which represents the set of paths that produce a given label prefix) and $\mathbb{S}$ (which represents the set of paths that produce a given label suffix), $\mathbb{C}$ represents the set of path-pairs (one path for each RNN) such that both paths produce the same sequence.

Probability that the Two RNNs Agree on a Particular Label Sequence. The probability that the two RNNs both generate label sequence ℓ, given that they agree, is

$$P(\ell|\mathbf{u},\mathbf{v}) = \frac{1}{Z}\alpha_{U,V}(\ell)$$

where

$$\alpha_{u,v}^{\epsilon}(\ell_{1:s}) = y_{\epsilon}^{u}\alpha_{u-1,v}(\ell_{1:s})$$
$$\alpha_{u,v}^{*\epsilon}(\ell_{1:s}) = z_{\epsilon}^{v}\alpha_{u,v-1}^{*}(\ell_{1:s})$$
$$\alpha_{u,v}^{**}(\ell_{1:s}) = y_{\ell_s}^{u}z_{\ell_s}^{v}\alpha_{u-1,v-1}(\ell_{1:s-1})$$
$$\alpha_{u,v}^{*}(\ell_{1:s}) = \alpha_{u,v}^{*\epsilon}(\ell_{1:s}) + \alpha_{u,v}^{**}(\ell_{1:s})$$
$$\alpha_{u,v}(\ell_{1:s}) = \alpha_{u,v}^{\epsilon}(\ell_{1:s}) + \alpha_{u,v}^{*}(\ell_{1:s})$$

with boundary condition $\alpha_{0,0}^{**}(\epsilon) = 1$ (and other "out-of-bounds" forward likelihoods defined to be zero).

The interpretation of these probabilities is

$$\alpha_{u,v}^{\epsilon}(\ell_{1:s}) = \sum_{\pi \in \mathbb{P}_u^{\epsilon}(\ell_{1:s})} P(\pi|\mathbf{u}) \sum_{\phi \in \mathbb{P}_v(\ell_{1:s})} P(\phi|\mathbf{v})$$

$$\alpha_{u,v}^{*\epsilon}(\ell_{1:s}) = \sum_{\pi \in \mathbb{P}_u^{*}(\ell_{1:s})} P(\pi|\mathbf{u}) \sum_{\phi \in \mathbb{P}_v^{\epsilon}(\ell_{1:s})} P(\phi|\mathbf{v})$$

$$\alpha_{u,v}^{**}(\ell_{1:s}) = \sum_{\pi \in \mathbb{P}_u^{*}(\ell_{1:s})} P(\pi|\mathbf{u}) \sum_{\phi \in \mathbb{P}_v^{*}(\ell_{1:s})} P(\phi|\mathbf{v})$$

Note that the label probability $P(\ell|\mathbf{u}, \mathbf{v})$ can, in fact, be calculated more efficiently by combining Eqs. 1 and 2, without reference to the pairwise Forward probabilities $\alpha_{u,v}^{**}(\kappa)$. However, the pairwise Forward probabilities are useful for calculating the prefix probability, $\sigma(\kappa|\mathbf{u}, \mathbf{v})$

$$\sigma(\kappa|\mathbf{u}, \mathbf{v}) = \frac{1}{Z} \sum_{(u,v) \in \mathbb{A}(\mathbf{u},\mathbf{v})} \alpha_{u,v}^{**}(\kappa)\gamma_{u+1,v+1} \qquad (3)$$

where $\mathbb{A}(\mathbf{u}, \mathbf{v}) = \{(u, v) : 1 \leq u \leq U, 1 \leq v \leq V\}$ is the *alignment envelope*.

Following [5,6], one general strategy for optimizing these calculations is to replace the exact expression for $\sigma(\kappa|\mathbf{u}, \mathbf{v})$ with a lower bound that is faster to compute, where the summation is constrained to some subset $\mathbb{A}'(\mathbf{u}, \mathbf{v}) \subset \mathbb{A}(\mathbf{u}, \mathbf{v})$ of the full alignment envelope, pre-identified as including the most probable alignment paths. After having sampled an alignment envelope $\mathbb{A}'$, one then computes the pairwise Forward probabilities $\alpha_{u,v}$, a computation that may be accelerated by visiting only cells with indices that are in the reduced envelope, $(u, v) \in \mathbb{A}'$. Finally, $\sigma(\kappa|\mathbf{u}, \mathbf{v})$ is computed as in Eq. 3, above.

Finding the Alignment Envelope. Having computed γ, an alignment envelope may be sampled by N iterations of stochastic trace (which samples from the set of paths where the two RNNs agree), as follows

- Initalize $\mathbb{A}' \leftarrow \emptyset$
- Repeat N times:
  - Initialize $(u, v) \leftarrow (0, 0)$

- While $u < U$ or $v < V$:
  * If $(u, v) \notin \mathbb{A}'$, then add (u, v) to $\mathbb{A}'$
  * Sample (u', v') with weight $\gamma_{u',v'}$ (or zero if $u' > U$ or $v' > V$) from the set $\{(u+1, v), (u, v+1), (u+1, v+1)\}$
  * Set $(u, v) \leftarrow (u', v')$

3   MCMC Consensus for Multiple Reads

We can alternatively formulate a complementary consensus approach using Markov chain Monte Carlo (MCMC). This has the advantage of being more easily generalized to more than two reads. This generic MCMC approach is similar to, and inspired by, Nanopolish [8].

Let $\mathbf{R} = \{(\mathbf{x}^{(i)}, \mathbf{y}^{(i)})\}$ denote a set of reads $\mathbf{x}^{(i)}$ and their associated RNN outputs $\mathbf{y}^{(i)}$. We augment this by an imputed state (to be sampled by MCMC) that specifies alignments of reads to an imputed consensus sequence: let $\mathbf{C} \in L^*$ denote the consensus and $\mathbf{A} = \{(\pi^{(i)}, S^{(i)})\}$ the alignments, where $\pi^{(i)}$ is the path through the i'th read's RNN. Let $\ell^{(i)} = \mathbb{B}(\pi^{(i)})$ denote the corresponding label sequence. We require at all times, and for all reads, that $\ell^{(i)}$ be a subsequence of $\mathbf{C}$, and that $S^{(i)}$ is the index (1-based with respect to $\mathbf{C}$) of the first label symbol $\ell_1^{(i)}$ for the i'th read. Thus, $E^{(i)} = S^{(i)} + |\ell^{(i)}| - 1$ is the index of the last label symbol, and $\ell^{(i)} = \mathbf{C}_{S^{(i)}:E^{(i)}}$.

We define a target distribution of the form

$$P(\mathbf{C}, \mathbf{A}|\mathbf{R}) \propto P(\mathbf{C}) \prod_i P(\ell^{(i)}|\mathbf{y}^{(i)})$$

where $P(\mathbf{C})$, a prior, penalizes sequences that get too long (e.g. IID with geometrically-distributed length).

One can readily describe a MCMC algorithm for sampling from this target distribution via

- Moves that resample the alignment of an individual read $(\pi^{(i)}, S^{(i)})$ while keeping the consensus sequence $\mathbf{C}$ fixed (and the other alignments)
  - Increase or decrease $S^{(i)}$ by removing or adding label symbols to the beginning of $\pi^{(i)}$
  - Sample directly from $P(\pi^{(i)}|\ell^{(i)} = \mathbf{C}_{S^{(i)}:E^{(i)}})$ using stochastic traceback through the $\alpha(\mathbf{C}_{S^{(i)}:E^{(i)}})$ matrix of Sect. 2.3
- Moves that resample the consensus sequence $\mathbf{C}$ while fixing the number and locations of gaps in the alignments
  - Insert base (adding a label character to the paths $\pi^{(i)}$ of any reads that overlap with that base)
  - Delete base (removing a label character from the paths of any overlapping reads)

The probabilities of these candidate moves $(\mathbf{C}', \mathbf{A}')$ can be calculated with $P(\pi^{(i)}|\mathbf{y}^{(i)})$, which leaves the Metropolis-Hastings acceptance probability

$$\min \left(1, \frac{P(\mathbf{C}', \mathbf{A}'|\mathbf{R})}{P(\mathbf{C}, \mathbf{A}|\mathbf{R})}\right)$$

where the proposal moves are accepted if the acceptance probability is greater than $u \in \mathrm{unif}(0, 1)$.

4   Connection to Finite State Automata

Since the RNN's output $\mathbf{y}$ is like the parameterization of a gapped position-specific weight matrix, we can represent it using a profile Hidden Markov Model, i.e. a state machine with (probabilistically) weighted transitions [3]. This is illustrated in Fig. 1. Denote this HMM by $\mathcal{R}(\mathbf{y})$.

Suppose ℓ represents a label sequence that may be the output of the RNN. One way to represent this constraint—that a sequence must have a particular value—is by using a second state machine as an indicator function (a one-hot vector for ℓ) that assigns unit weight to sequence ℓ, and zero weight to anything else. Denote this second state machine by $\mathcal{S}(\ell)$.

To impose the constraint that two weighted automata $\mathcal{A}$ and $\mathcal{B}$ emit the same sequence, we use a well-defined algorithm [11] to take the intersection $\mathcal{AB}$. That is, we construct a combined state machine whose individual paths represent alignments of paths through $\mathcal{A}$ and $\mathcal{B}$ emitting the same thing. The combined machine uses a product of $\mathcal{A}$ and $\mathcal{B}$'s transition graphs, as illustrated for $\mathcal{R}(\mathbf{y})\mathcal{S}(\ell)$ in Fig. 3.

Several quantities of interest are then well-defined in terms of the norms (i.e. sum-over-all-paths) of these intersected machines, including:

- The probability $P(\ell|\mathbf{y})$ that the RNN's output decodes to sequence ℓ is $|\mathcal{R}(\mathbf{y})\mathcal{S}(\ell)|$ and this probability takes time $\mathcal{O}(T|\ell|)$ to compute
- The probability $\sigma(\kappa|\mathbf{y})$ that it decodes to a sequence whose prefix is κ is $|\mathcal{R}(\mathbf{y})\mathcal{P}(\kappa)|$, where $\mathcal{P}(\kappa)$ is the indicator machine matching any sequence beginning with κ
- The probability Z that two RNNs with outputs $\mathbf{y}, \mathbf{z}$ decode to the same sequence is $|\mathcal{G}|$ where $\mathcal{G} = \mathcal{R}(\mathbf{y})\mathcal{R}(\mathbf{z})$
- The probability $P(\ell|\mathbf{y}, \mathbf{z})$ that ℓ is the consensus decoded sequence is $\frac{1}{Z}|\mathcal{G}\mathcal{S}(\ell)| = \frac{1}{Z}|\mathcal{R}(\mathbf{y})\mathcal{S}(\ell)| \cdot |\mathcal{R}(\mathbf{z})\mathcal{S}(\ell)|$; the probability $\sigma(\kappa|\mathbf{y}, \mathbf{z})$ that κ is its prefix is $\frac{1}{Z}|\mathcal{G}\mathcal{P}(\kappa)|$; and so on
- Calculations involving $\mathcal{G}$ take time and memory $\mathcal{O}(UV)$, which is expensive. However, optimized approximations are possible, e.g. using a reduced-size machine $\mathcal{G}' \subset \mathcal{G}$

The states of $\mathcal{G}$ and $\mathcal{G}\mathcal{S}(\ell)$ are related to the recursions for $\gamma_{u,v}^{\epsilon}, \gamma_{u,v}^{*\epsilon}, \gamma_{u,v}^{**}$ and $\alpha_{u,v}^{\epsilon}, \alpha_{u,v}^{*\epsilon}, \alpha_{u,v}^{**}$ in Sect. 2.4. Restricting the alignment envelope to a subset $\mathbb{A}'$ of the full alignment envelope $\mathbb{A}$ is equivalent to using a smaller automaton $\mathcal{G}'$ whose transition graph is a subgraph of the full transition graph of $\mathcal{G}$.

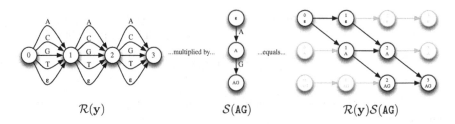

$$\mathcal{R}(\mathbf{y}) \qquad \mathcal{S}(\text{AG}) \qquad \mathcal{R}(\mathbf{y})\mathcal{S}(\text{AG})$$

Fig. 3. Intersection of automata (analogous to pointwise vector multiplication) constrains two machines to emit the same thing. The intersection shown here is written $\mathcal{R}(\mathbf{y})\mathcal{S}(\text{AG})$ where $\mathcal{R}(\mathbf{y})$ represents the machine on the left, derived from a neural network (see Fig. 1) and $\mathcal{S}(\text{AG})$ represents the machine in the middle, which allows only the sequence AG. Intersection is implemented by taking the product of the two transition graphs, keeping only edges where the labels of the two machines are synchronized, with an arbitrary well-defined ordering on gaps [1,7,9,11]. Some states and transitions in the intersected machine may be inaccessible (shown grayed-out).

Further details may be found in a supplement available on the paper's website (https://jordisr.github.io/consensus-rnns/) and in references [1,7,9,11].

Acknowledgments. The authors were supported by NIH/NCI grant CA220441 and by NIH/NHGRI training grant T32 HG000047. We thank the anonymous reviewers for their helpful comments.

References

1. Bouchard-Côté, A.: A note on probabilistic models over strings: the linear algebra approach. Bull. Math. Biol. **75**(12), 2529–2550 (2013)
2. David, M., Dursi, L.J., Yao, D., Boutros, P.C., Simpson, J.T.: Nanocall: an open source basecaller for Oxford nanopore sequencing data. Bioinformatics **33**(1), 49–55 (2017)
3. Durbin, R., Eddy, S., Krogh, A., Mitchison, G.: Biological Sequence Analysis: Probabilistic Models of Proteins and Nucleic Acids. Cambridge University Press, Cambridge (1998)
4. Graves, A., Fernández, S., Gomez, F., Schmidhuber, J.: Connectionist temporal classification: labelling unsegmented sequence data with recurrent neural networks. In: Proceedings of the 23rd International Conference on Machine Learning, ICML 2006, pp. 369–376. ACM, New York (2006). https://doi.org/10.1145/1143844.1143891
5. Holmes, I.: Accelerated probabilistic inference of RNA structure evolution. BMC Bioinform. **6**(73) (2005)
6. Holmes, I., Durbin, R.: Dynamic programming alignment accuracy. J. Comput. Biol. **5**(3), 493–504 (1998)
7. Holmes, I.H.: Historian: accurate reconstruction of ancestral sequences and evolutionary rates. Bioinformatics **33**(8), 1227–1229 (2017)
8. Loman, N.J., Quick, J., Simpson, J.T.: A complete bacterial genome assembled de novo using only nanopore sequencing data. Nat. Methods **12**(8), 733–735 (2015)

 9. Mohri, M., Pereira, F., Riley, M.: Weighted finite-state transducers in speech recognition. Comput. Speech Lang. **16**(1), 69–88 (2002)
10. Teng, H., Hall, M.B., Duarte, T., Cao, M.D., Coin, L.: Chiron: Translating nanopore raw signal directly into nucleotide sequence using deep learning. bioRxiv (2017). https://doi.org/10.1101/179531, https://www.biorxiv.org/content/early/2017/08/23/179531
11. Westesson, O., Lunter, G., Paten, B., Holmes, I.: Accurate reconstruction of insertion-deletion histories by statistical phylogenetics. PLoS One **7**(4), e34572 (2012)

Utilize Imputation Method and Meta-analysis to Identify DNA-Methylation-Mediated microRNAs in Ovarian Cancer

Ezra B. Wijaya[1], Erwandy Lim[1], David Agustriawan[2],
Chien-Hung Huang[3], Jeffrey J. P. Tsai[1], and Ka-Lok Ng[1,4(✉)] ⓘ

[1] Department of Bioinformatics and Medical Engineering,
Asia University, Taichung, Taiwan
ppiddi@gmail.com
[2] Department of Bioinformatics,
Indonesia International Institute for Life Sciences, East Jakarta, Indonesia
[3] Department of Computer Science and Information Engineering,
National Formosa University, Huwei, Taiwan
[4] Department of Medical Research, China Medical University Hospital,
China Medical University, Taichung, Taiwan

Abstract. Both epigenetics and genetic alterations are associated with cancer formation. Identification of prognostic biomarkers, DNA-methylation-mediated miRNAs, is an important step towards developing therapeutic treatment of cancer. Ovarian cancer is one of the most lethal cancers among the females, it was selected for the present study. The TCGA database provides large volume of data for cancer study, which is useful if one can combine different batches of datasets; hence, higher confident results can be obtained. There are several issues arise in data integration, i.e., missing data problem, data heterogeneity problem and the need of construct an automatic platform to reduce human intervention.
Method. Both the normal and ovarian tumor datasets were obtained from the TCGA database. To interpolate the missing methylation values, we employed the *KNN* imputation method. Simulation tests were performed to obtain the optimal *k* value. We utilized meta-analysis to minimize the heterogeneity problem and derived statistical significant DNA methylation-mediated-miRNA events. Finally, a semi-automatic pipeline was constructed to facilitate the imputation and meta-analysis studies; thus, identify potential epigenetic biomarkers in a more efficient manner.
Results. Both epigenetic- and TF-mediated effects were examined, which allow us to remove false positive events. The methylation-mediated-miRNA pairs identified by our platform are in-line with literature studies.
Conclusion. We have demonstrated that our imputation and meta-analysis pipeline led to better performance and efficiency in detecting methylation-mediated-miRNA pairs. Furthermore, this study reveals the association between aberrant DNA methylation and alternated miRNA expression, which contributes to better knowledge of the role of epigenetics regulation in ovarian cancer formation.

E.B. Wijaya and E. Lim—Equal contribution.

© Springer International Publishing AG, part of Springer Nature 2018
J. Jansson et al. (Eds.): AlCoB 2018, LNBI 10849, pp. 140–153, 2018.
https://doi.org/10.1007/978-3-319-91938-6_12

Keywords: Imputation · Ovarian cancer · DNA methylation · microRNA
Meta-analysis · Transcription factor

1 Introduction

Epigenomics study not only provides a normal functional process in cell differentiation but also may reveal the relations between methylation patterns and cancer diseases [1]. For example, aberrant methylation in CpG island shores and gene bodies may affect large genomic regions in colorectal cancer [2]. Similarly, histone methylation and acetylation is altered in breast cancer cell [3]. Moreover, aberrant expression of miRNA, another regulatory element of epigenetics disease, shows responsible in gastric cancer angiogenesis [4].

Ovarian cancer is one of the most dangerous cancers among other cancers of the female reproductive system that causes higher mortality rate in women. In 2013, there were 20,927 women in the United States were diagnosed with ovarian cancer and 14,276 died because of this cancer [5]. Furthermore, ovarian cancer is one of the diseases that has been studied both in genetics and epigenetics. Ovarian serous cystadenocarcinoma (OSC) was chosen for the present study.

miRNAs can contribute disease progression by acting as oncogenes (OCGs) or tumor suppressor genes (TSGs) [6] and also negatively regulate genes involved in cell proliferation and cell survival which cause the instability of gene transcription and protein development [7]. Aberrant DNA methylation has been considered as one of the epigenetic mechanisms that controls both up- and down-regulation of miRNA expression, which led to cancer formation [8].

Multiples studies have shown the interaction between methylation and miRNA expression in cancer progression. For example, down regulation of miR-10b* in breast tumors due to hypermethylation of CpG islands located upstream promotes tumor proliferation [9, 10]. Additionally, recent study has shown hypermethylation of miR-193a-3p resulted in *GRB7* upregulation which enhance oncogenic properties of ovarian cancer cells *in vitro* and *in vivo* [11].

Moreover, experimental studies show that transcription factors (TFs) and miRNAs can regulate each other. Delfino et al. [12] identified eight TFs involve in ovarian cancer; such as, circadian locomotor output cycles kaput (CLOCK), estrogen receptor 2 (ESR2), v-etsery throblastosis virus E26 oncogene homolog 2 (ETS2), histone deacetylase 3 (HDAC3), homeobox A1 (HOXA1), v-myc myelocytomatosis viral oncogene homolog (MYC), nuclear receptor subfamily 5, group A, member 1 (NR5A1), and POU class 2 homeobox 2 (POU2F2).

In recent years, a large amount of information on ovarian cancer study is well documented in The Cancer Genome Atlas (TCGA) database, a resource provided by the National Cancer Institute and the National Human Genome Research Institute [13, 14]. Despite TCGA delivers massive genomic information, some of the microarray and NGS data might be missing while making measurements. As missing data can reduce the accuracy in the results, a study [15] shown that *KNN* imputation can be used as a standard to interpolate the missing values due to its accuracy performance. It has shown that only 6–26% average deviation from the true values of the estimated values and the error rate is under 0.25 after normalization [15, 16]. On the other hand, meta-analysis could provide a rigorous statistical framework for summarizing the results of multiple experiments in a single estimate [17].

In our previous work [8], we utilized meta-analysis to identify DNA-methylation-mediated miRNA in OSC using 12 batches of datasets. In this study, we extended our previous work by performing five additional tasks: (i) determine the optimal k value by conducting missing data simulation experiments, (ii) make use of the imputation method to infer missing methylation values, (iii) quantify the difference between the normal and tumor cases using two different parameters: i.e. the methylation levels (so-called beta-value or β-value) and the *log₂(fold change)*, (iv) apply *Student's t-test* to infer aberrant methylation and differentially expressed miRNAs, and (v) develop a semi-automatic pipeline to facilitate data integration, data pre-processing, and meta-analysis study for 14 batches of data, i.e. two more batches compare to our previous work [8].

2 Methods

2.1 Dataset

We obtained 14 batches of tumor samples (569 cases) and one batch of normal sample (12 cases) of both DNA methylation and miRNA expression data from TCGA. DNA methylation profiles were measured by using the Illumina Infinium human DNA methylation 27 platform (consists of 27578 DNA methylation sites), which provided methylation probe ID, chromosome locations, genomic locations and each patient's CpG islands' methylation levels. The miRNA expression levels were recorded by using the Agilent 8x15K Human miRNA-specific microarray platform (consists of 799 miRNAs), which provided miRNA ID and miRNA expression score information. All the beta-values and miRNA expression information used in this study are normalized data ('Level 3' data).

2.2 Construction of a Semi-automatic Pipeline

We developed a semi-automatic pipeline to accelerate the data processing and analysis process. This pipeline provides the following functionalities: (i) it integrates DNA methylation and miRNA expression datasets, (ii) it interpolates missing methylation values using the *KNN* imputation method, and (iii) it performs the heterogeneity test and meta-analysis study. Figure 1 showed the workflow of our work.

Prior to pipeline construction, we designed a folder management process to organize the methylation and miRNA expression data. The folder management organization structure was provided in Fig. 2.

KNN Imputation. After combined the batch information, we handled the missing data problem. This process replaces the missing values with some values resemble the individual test value or approximately has the same distribution [18]. For probe x_b with missing beta-value, its methylation level j is estimated by taking the average of the k nearest probes' methylation levels $x_{\alpha i}$. The missing beta-value of probe x_{bj} is given by,

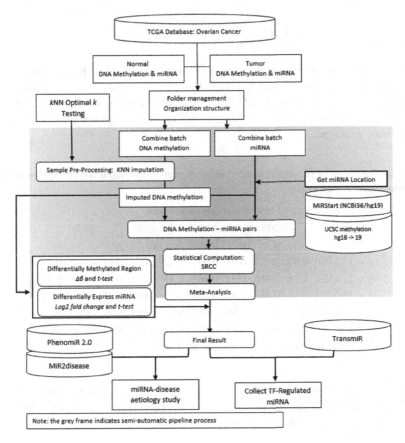

Fig. 1. The workflow of utilizing imputation and meta-analysis DNA-methylation-mediated miRNA pipeline

$$x_{bj} = \frac{\sum_{\forall x_\alpha \in N_b} x_{\alpha i}}{k} \tag{1}$$

where N_b is the set of k nearest neighbors of probe x_b.

To determine the optimal k, we conducted simulation test to find the best performing k value. A study reported by Ahmadi-Nedushan [19], suggested optimal k lies between 1, 2, 3, 4 or 5. We used the 'batch 40' dataset (from TCGA) as our control, because: (i) it consists of the largest number of cases, i.e. 49, and (ii) it has no missing value. Simulated datasets were generated by randomly assigned five different levels, i.e. 10%, 20%, 30%, 40%, and 50%, of missing values for each row in the control dataset. Then we applied the *KNN* method, using five different k, i.e. 1 to 5, to impute the missing values.

To infer the optimal k, both statistical *t-test* analysis and *matrix difference calculation* were conducted. The difference between matrices A and B is given by:

$$d(\boldsymbol{A}, \boldsymbol{B}) = \sqrt{\sum_{i=1}^{n} \sum_{j=1}^{n} \left(a_{ij} - b_{ij}\right)^2} \tag{2}$$

where A = the control matrix, B = the imputated matrix, a_{ij} = the ith row and jth column matrix element of A, and b_{ij} = the ith row and jth column matrix element of B. All analysis was done using MATLAB® (R2016b) function $knnimpute()$ and t-test analysis using the function $ttest2()$. Performance tests were repeated five times and the results were averaged.

Table 1. The results of the performance of *KNN* imputation in datasets with five different levels of randomized missing values

KNN	Percentage of randomized missing values									
	10%		20%		30%		40%		50%	
	d	p_t	d	p_t	d	p_t	d	p_t	d	p_t
$k = 1$	0.060	0.97	0.071	0.91	0.071	0.84	0.071	0.83	0.071	0.77
$k = 2$	0.056	0.96	0.066	0.92	0.067	0.87	0.068	0.86	0.068	0.81
$k = 3$	0.053	0.96	0.063	0.94	0.065	0.86	0.066	0.83	0.066	0.78
$k = 4$	0.053	0.96	0.061	0.96	0.063	0.89	0.064	0.87	0.064	0.82
$k = 5$	0.052	0.96	0.059	0.97	0.062	0.92	0.063	0.92	0.063	0.87

d = matrix difference; p_t = p-value of the *t-test*

As shown in Table 1, k equals to 5 shows the best performance which has the smallest matrix difference. The results of the *t-test* suggests that there is no significant difference; i.e. $p_t > 0.05$, between the control and the imputed matrix.

Meta-analysis Study. We examined **direct** DNA-methylation-mediated events by sorting out methylation sites that are located within 100 kb (both upstream and downstream) of a miRNA's TSS [20]. The genomic coordinates of the DNA methylation sites are based on NCBI36/hg18, while the TSS of the miRNAs' genomic positions are based on GRCh37/hg19 from MiRStart [21] and MiRBase [22]. To adjust the difference between two human genome versions, we employed the 'Lift Over' tool, which is provided by the UCSC genome browser, to map the methylation sites' coordinates to the GRCh36/hg18 version. Then we performed the Spearman's Rank Correlation Coefficient (*SRCC*) calculation to estimate the correlation strength between the methylation level and the miRNA expression level. To conduct the meta-analysis study, the *SRCC* were first transformed into the Fisher's Z metric, and all the calculations were conducted using the Z-values [23].

As proposed by Agustriawan et al. [8], we handled the heterogeneous problem by performing the I^2 statistical test, where I^2 is defined by [20],

$$I^2 = \frac{Q - df}{Q} \times 100\% \tag{3}$$

where df denotes the degree of freedom, and Q is given by,

$$Q = \sum_{p=1}^{q} W_p \left(Y_p - M \right)^2 \tag{4}$$

where q, W, Y, M represent the number of studies, the study weight, the study effect size and the summary effect, respectively [23]. This test assessing the dispersion across batches with 25%, 50% and 75% heterogeneity values might be interpreted as low, moderate and high variances, respectively. I^2 also can be interpreted as the signal to noise ratio between batches, so if I^2 shows statistically significant result, we can conclude that the results across batches do not share a common effect size, i.e. the Z-values or SRCC, used in the present study. At the end of the meta-analysis study, the Z-values were transformed back to the SRCC. Furthermore, the confidence intervals of SRCC and meta-analysis p-values, p_{MA}, were estimated.

Differential Expression Analysis of DNA Methylation and miRNA Expression. We conducted Differential Methylation Region (DMR) study and differential expressed miRNA analysis. In DMR study, we applied the difference in mean β-value concept,

$$\Delta\beta = (\text{mean } \beta\text{-value in tumor case}) - (\text{mean } \beta\text{-value in normal case}) \tag{5}$$

As suggested by Sung et al. [24], a methylation site is hypermethylated or hypomethylated if $\Delta\beta$ was greater than 0.2 or less than −0.2, respectively. In order to have better knowledge of the mean differences, *t-test* analysis was applied using the MATLAB® package *mattest()* with VARTYPE set as 'unequal' for unknown variance. Methylation level is considered as hypermethylated if $\Delta\beta > 0.2$ and hypomethylated if $\Delta\beta < -0.2$ with *t-test* shows statistically significant ($\alpha = 0.05$).

To determine differentially expressed miRNAs, we compared the tumor and normal cases by using the two-sample *t-test*. If $p_t < 0.05$, then the difference is significant. Negative $log_2(fold\ change)$ represents downregulation while positive logarithmic value represents upregulation.

2.3   Comparison of DNA-Methylation-Mediated miRNA Predictions with Disease-Related miRNA Databases

Aside from epigenetics alterations, which play a role in regulating the miRNAs in cancer study, transcription factor (TFs) may also control gene expression by activating or deactivating the DNA transcription process. To the best our knowledge, there are very limited works addressing the interaction between these two mechanisms. We examined this by utilizing information retrieved from a TF-mediated-miRNA database, TransmiR (August 7th, 2016 version) [25]. Also, to validate the methylation-mediated-miRNA results, we compared our predicted miRNAs with both the Pheno-miR 2.0 [26] and the MiR2Disease [27] databases.

3 Results

3.1 KNN Imputation Results

Given the Illumina platform, we scanned a total of 27578 methylation probes and found 2603 rows (methylation probes) with >50% missing values. After removed methylation probes with >50% missing values, we applied *KNN* imputation with $k = 5$ to interpolate the missing values for the 24975 remaining probes. As a result, there are a minimum of 24 probes and a maximum of 2569 probes' beta-values were imputed. Furthermore, the number of predicted methylation-mediated miRNA pairs increased from 1718 without imputation to 2512 with imputation. The number of DNA-methylation-mediated miRNAs was reduced from 15 significant pairs using imputation to nine without imputation due to increase of noise ratio ($I^2 < 0.05$, data not shown). Among the nine events, two miRNAs are related to ovarian cancer. Four events are common to both studies. Imputation has the potential in detecting more methylation-mediated-miRNA pairs; however, the method may result in removing a few real events. It is not clear whether imputation will incur such effect in general or not, more studies are need to settle this question.

3.2 The Results of Meta-analysis Study and TF-Mediated-miRNAs

By comparing the genomic locations of the methylation sites and the miRNA TSS, we found 2512 pairs, which include 1118 methylation sites and 375 miRNAs. Some of the miRNAs may have two or more methylation regions falling within 100 kbps of the miRNA TSS. Given the beta-values and the miRNA expression levels, the *SRCC* were calculated. Next, our pipeline performed the heterogeneity test and the meta-analysis calculation. Heterogeneity test showed that there is no significant variance across batches. The test shows that all the 14 batches share the common effect size, *SRCC*. Then, the final *SRCC* and p-values (p_{MA}) for the 2512 pairs were calculated. It was found that 586 pairs are significant, i.e. $p_{MA} < 0.05$, Out of the 586 pairs, 376 pairs are anti-correlation events.

To access aberrant alternations, both DMR study and differential expressed miRNA study were performed. Out of the 376 events, we found 11 methylation sites are differentially methylated and 13 miRNAs are differentially expressed (Table 2). We noted that among the 11 methylation sites, hypomethylation is the dominant regulation mechanism, which is consistent with the general consensus that global hypomethylation occur in cancer [28] and in ovarian cancer [29]. Moreover, we observed that the hypomethylated probe, cg12019109, mediates upregulation of hsa-miR-25*. This event has the most negative *SRCC*, i.e. $r = -0.46$ and $p_{MA} \ll 0.05$.

It is known that [30] the interaction between TF and methylation can be mediated through DNA methylation and chromatin interaction. Methylation of DNA results in a chromatin structure that may prevent TF from accessing DNA. As a result, the tran-

Table 2. Significant correlated DNA-methylation-mediated miRNAs

Methylation site	Methylation status	miRNA	miRNA expression	SRCC	p_{MA}
cg12019109	Hypomethylated	hsa-miR-25*	Upregulated	−0.46	0.00
cg07446572	Hypomethylated	hsa-miR-200c	Upregulated	−0.30	1.16E−12
cg07446572	Hypomethylated	hsa-miR-200c*	Upregulated	−0.30	1.83E−12
cg07109801	Hypomethylated	hsa-miR-425	Upregulated	−0.27	1.18E−10
<u>cg07297906</u>	Hypomethylated	hsa-miR-20b	Upregulated	−0.21	6.41E−07
cg11063110	Hypomethylated	hsa-miR-335	Upregulated	−0.20	2.54E−06
cg06641503	Hypomethylated	hsa-miR-191	Upregulated	−0.20	5.38E−06
cg12019109	Hypomethylated	hsa-miR-93	Upregulated	−0.18	2.32E−05
<u>cg07297906</u>	Hypomethylated	hsa-miR-106a	Upregulated	−0.18	2.95E−05
cg10722799	Hypermethylated	hsa-miR-802	Downregulated	−0.14	0.001
cg21578541	Hypomethylated	hsa-let-7g	Upregulated	−0.13	0.002
cg08077673	Hypomethylated	hsa-miR-29a*	Upregulated	−0.13	0.002
cg01888566	Hypomethylated	hsa-miR-29a*	Upregulated	−0.13	0.002
cg12019109	Hypomethylated	hsa-miR-106b*	Upregulated	−0.10	0.021
cg13118849	Hypomethylated	hsa-miR-640	Upregulated	−0.09	0.039

p_{MA} denotes the p-value for the meta-analysis study, bold, italic and underline fonts denote the same methylation sties.

scription will not occur. In contrast, unmethylated DNA involved in chromatin remodeling and chromatin modification which permit the TF hybridizes with the DNA.

Among the 15 events, 14 significant DNA-methylation-mediated miRNAs (hsa-miR-29a* appear twice) were identified, where four miRNAs are regulated by TFs (Table 3). If both epigenetic and TF regulations appear to associate miRNA expression; it suggests that the DNA-methylation-mediated prediction may be a false positive event. For example, *E2F1* and *PTEN* act as an activator of miRNA-25*. From Table 2 results, miRNA-25* is associated with a hypomethylated site, thus, we can assume that hypomethylated miRNA-25* is activated by *E2F1* and *PTEN*. As we shown in Table 4, miR-25* does involve in ovarian cancer formation, which is recorded by both the PhenomiR and the MiR2Disease databases.

MiR-93 is regulated by three TFs, where *E2F1* acts as a regulator, *PDGF-B* and *MYC* act as suppressors. If miR-93 is hypomethylated, then it will be down-regulated by *PDGF-B* or *MYC*, which is not consistent with the hypomethylate-mediate effect; hence, there may be other effects involved. Similarly, *E2F1* acts as a regulator of miRNA-106b; and miR-29a* is regulated by nine TFs, i.e. *HMGA1, MYC, NFKB1, YY1, IL-4, PDGF-B, TGFB1, CEBPA* and *AP-1*. Among the nine TFs, two TFs activate miRNA-29* expression. Again, both miR-106b and miR-29a* may involve other effects beside DNA-methylation.

Table 3. TransmiR - ovarian cancer associated TF-mediated miRNAs

TF	TF-regulated miRNA	Action	miRNA expression	Methylation probes
E2F1	miR-25*	Regulation	Upregulated	cg12019109
PTEN	miR-25*	Activation	Upregulated	cg12019109
E2F1	miR-93	Regulation	Upregulated	cg12019109
PDGF-B, MYC	miR-93	Repression	Upregulated	cg12019109
HMGA1	miR-29a*	Activation	Upregulated	cg08077673, cg01888566
MYC	miR-29a*	Repression	Upregulated	cg08077673, cg01888566
NFKB1	miR-29a*	Repression	Upregulated	cg08077673, cg01888566
YY1	miR-29a*	Repression	Upregulated	cg08077673, cg01888566
IL-4	miR-29a*	Repression	Upregulated	cg08077673, cg01888566
PDGF-B	miR-29a*	Repression	Upregulated	cg08077673, cg01888566
TGFB1	miR-29a*	Repression	Upregulated	cg08077673, cg01888566
CEBPA	miR-29a*	Repression	Upregulated	cg08077673, cg01888566
AP-1	miR-29a*	Activation	Upregulated	cg08077673, cg01888566
E2F1	miR-106b*	Regulation	Upregulated	cg12019109

3.3   DNA-Methylation-Mediated miRNAs and Disease-Related Databases

From the PhenomiR and MiR2disease database records, it was found that there are 10 miRNAs (from Table 4, hsa-miR-25*, hsa-miR-200c, hsa-miR-335, hsa-miR-191, hsa-miR-93, hsa-miR-106a, hsa-let-7g, hsa-miR-29a*, hsa-miR-106b*, hsa-miR-640) consistently upregulated and are involved in ovarian cancer. Each one of these miR-NAs is associated with a hypomethylated site (Table 2). Our study suggested that due to aberrant methylation, those miRNAs are associated with ovarian cancer formation. In particular, among these 10 miRNAs, four of them (Table 4) were recorded upregulated in our prediction as well as in the PhenomiR or the MiR2Disease database. According to PhenomiR and MiR2Disease, there are four miRNAs (hsa-let-7g, hsa-miR-106b*, hsa-miR-29*, and hsa-miR-335) were reported to be both up- and down-regulated. Also, the rest three miRNAs (hsa-miR-20b, hsa-miR-425, and hsa-miR-802) remain unrecorded in the literature. Therefore, further study is required to clarify this confusing situation.

Table 4. The results miRNAs prediction involved in ovarian cancer according to Phenomir and OncomirDB

miRNA	Probe	miRNA regulation/effect			References (PMID)
		P	PhR	M2D	
let-7g	cg21578541	Up	Up	Down	18442408, 17600087
miRNA-106a	cg07297906	Up	Up	–	18442408
miRNA-106b*	cg12019109	Up	Up/down	Down	16754881, 18560586, 18442408
miRNA-191	cg06641503	Up	–	–	27419385, 26191186, 25819812, 20167074
miRNA-200c	cg07446572	Up	Up	Up	17875710, 18451233, 17875710, 19435871, 19854497
miRNA-20b	cg07297906	Up	–	–	–
miRNA-25*	cg12019109	Up	Up	–	16754881
miRNA-29a*	cg08077673/ cg01888566	Up	Up/down	Up	16754881, 18560586, 17875710, 18199536, 18451233
miRNA-335	cg11063110	Up	Up/down	Down	16754881, 18560586
miRNA-425	cg07109801	Up	–	–	–
miRNA-640	cg13118849	Up	–	–	23627607
miRNA-802	cg10722799	Down	–	–	–
miRNA-93	cg12019109	Up	Up	Up	16754881, 18451233, 18442408

P = Prediction; PhR = PhenomiR; M2D = MiR2Disease. Note 1: the * symbol denotes the less predominant miRNA derives from the same hairpin of the pre-miRNA; 3p and 5p denotes the direction of miRNA is from 3' arm and 5' arm, respectively. The sequence of the miRNA is predominant. Note 2: PhenomiR (PhR) and MiR2Disease (M2D) database might cite using the same references PMI.

4   Discussion and Conclusion

Although epigenetics has very complex mechanisms and many factors involved leading to ovarian cancer, we investigated *direct* DNA-methylation-mediated miRNA expression in this work. Both DNA-methylation and TFs could cause abnormal miRNA expression, where aberrant DNA-methylation level may be one of the mechanisms that induce the altered miRNA expression. It is possible that certain TFs may interfere epigenetic alterations. However, those miRNAs are putative epigenetic biomarkers, further study is needed to investigate their involvement in ovarian cancer.

Compared with two cancer-associated miRNA databases, we identified 10 miRNAs exhibit aberrant alterations in ovarian cancer (Table 4). In Table 4, some miRNAs reported as being both up and down-regulated. The conflicting reports can be occurred because of the similar or dissimilar samples (e.g. real tumours vs cell lines, etc.), unbalance number of samples between cancer and normal patients, lack of normal samples can be a limitation to identify differentially expressed miRNAs.

Some of miRNA such as hsa-let-7g might have different results due to sample differences in their study design. PhenomiR database reported patients study phenotype [31] while in Mir2disease database reporting miRNA expression using Type I/SC1 and

Type II/SC2 cell lines model [32]. Similar to hsa-let-7g, hsa-miRNA-106b* has inconsistency result due to multiple study designs reported especially in PhenomiR database while another miRNA remains unrecorded in PhenomiR or Mir2disease database. Furthermore, according to meta-analysis result these miRNAs are significantly associated with at least one hypomethylated region, suggesting that they are potential epigenetic biomarkers in ovarian cancer.

There are some limitations of the present work, that is, TCGA provided methylation data measured by the Illumina InfiniumHumanMethylation27 platform instead of Infinium, HumanMethylation450 Bead Chip array (January 15, 2016). This platform provide data for more than 485,000 methylation sites at single-nucleotide resolution. Unfortunately, TCGA provided a few patients' 450k data, hence subjected to large statistical variation problem. If more samples are available in the future, we can apply our computation pipeline to conduct the same analysis. Moreover, the published literature did not provide epigenetic-mediated miRNA information for ovarian cancer; therefore, it is difficult to evaluate the accuracy of our predictions.

In conclusion, this work demonstrated a comprehensive analysis strategy to predict **direct** DNA-methylation-mediated miRNAs in OSC. The *KNN* imputation method was utilized to interpolate the missing β-values using an optimal k value determined by simulation experiments. Data heterogeneity and multi-batches issues are solved by performing meta-analysis study. Both normal and ovarian tumor cases were employed to determine the correlation between aberrant DNA methylation and alternated miRNA expression.

A pipeline has been developed to facilitate the above steps in a semi-automatic manner. This pipeline can be easily extended to analyze large-scale data for other cancers, i.e. pan-cancer studies. Also, we have removed TF-mediated miRNAs, thus, highly confident epigenetic-regulated miRNA events were obtained. The effectiveness of this pipeline is supported by the literature.

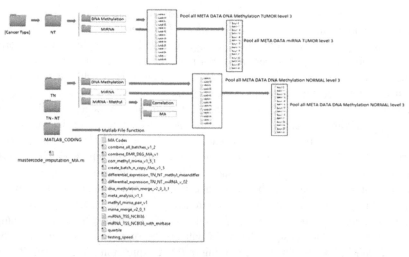

Fig. 2. Folder management for organizing data

We created a root folder named *'Cancer Type'* with four subfolders name: (i) normal sample data (NT), (ii) tumor sample data (TN) (iii) both normal and tumor datasets (TN – NT) for differentially expression analysis and (iv) MATLAB_CODING folder for collecting MATLAB programs.

NT folder, consist of normal sample DNA methylation and miRNA batch. This dataset manually collected and move according to their batch number folder.

TN folder, similar to *NT folder*, tumor sample of DNA methylation and miRNA batch dataset and collected according to its batch number. Add MiRNA-Methyl subfolder which consist of *Correlation* and *MA* folder (directory output for correlation and meta-analysis statistical result).

TN-NT folder, final output directory location of Differentially Methylated Region (DMR), differentially express miRNA, and meta-analysis with methylation-miRNA pair expression analysis.

MATLAB_CODING folder, all MATLAB programs for data integration, imputation, correlation analysis, meta-analysis, and expression analysis. All MATLAB function coding inside this folder will be executed through *mastercode_imputation_MA.m* function.

mastercode_imputation_MA.m function, an executable MATLAB coding to automatically process all analysis function collected in the MATLAB_CODING folder.

Acknowledgments. Dr. Chien-Hung Huang work is supported by the Ministry of Science and Technology (MOST) under the grant of MOST 106-2221-E-150-056. Dr. Ka-Lok Ng work is supported by the grants of MOST 106-2221-E-468-017, MOST 106-2632-E-468-002, and also supported under the grant from Asia University, 105-asia-1, and 106-asia-06. Dr. Jeffrey J. P. Tsai work is supported by the grant of MOST 106-2632-E-468-002. Mr. Ezra B. Wijaya work is supported by the grant of MOST 106-2221-E-468 -017.

References

1. Feinberg, A.P.: Genome-scale approaches to the epigenetics of common human disease. Virchows Arch. Int. J. Pathol. **456**(1), 13–21 (2010). https://doi.org/10.1007/s00428-009-0847-2
2. Jia, Y., Guo, M.: Epigenetic changes in colorectal cancer. Chin. J. Cancer **32**(1), 21–30 (2013). https://doi.org/10.5732/cjc.011.10245
3. Monteiro, F.L., Vitorino, R., Wang, J., et al.: The histone H2A isoform Hist2h2ac is a novel regulator of proliferation and epithelial–mesenchymal transition in mammary epithelial and in breast cancer cells. Cancer Lett. **396**, 42–52 (2017). https://doi.org/10.1016/j.canlet.2017.03.007
4. Zhang, X., Dong, J., He, Y., et al.: miR-218 inhibited tumor angiogenesis by targeting ROBO1 in gastric cancer. Gene **615**, 42–49 (2017)
5. U.S. Cancer Statistics Working Group: United States Cancer Statistics: 1999–2013 Incidence and Mortality Web-based Report. U.S. Department of Health and Human Services, Centers for Disease Control and Prevention and National Cancer Institute, Atlanta, 11 February 2017. www.cdc.gov/uscs

6. Lujambio, A., Ropero, S., Ballestar, E., et al.: Genetic unmasking of an epigenetically silenced microRNA in human cancer cells. Cancer Res. **67**, 1424–1429 (2007). https://doi.org/10.1158/0008-5472.can-06-4218

7. Monroig, P.C., Calin, G.A.: MicroRNA and epigenetics: diagnostic and therapeutic opportunities. Curr. Pathobiol. Rep. **1**(1), 43–52 (2013)

8. Agustriawan, D., Huang, C.-H., Sheu, J.J.-C., et al.: DNA methylation-mediatedmicroRNA pathways in ovarian serous cystadenocarcinoma: a meta-analysis. Comput. Biol. Chem. **65**, 154–164 (2016)

9. Biagioni, F., Ben-Noshe, N.B., Fontemaggi, G., et al.: miR-10b, a master inhibitor of the cell cycle, is down-regulated in human breast tumours. EMBO Mol. Med. **4**(11), 1214–1229 (2012). https://doi.org/10.1002/emmm.201201483

10. Biagioni, F., Bossel, B.-M.N., Fontemaggi, G., et al.: The locus of microRNA-10b: a critical target for breast cancer insurgence and dissemination. Cell Cycle **12**(15), 2371–2375 (2013). https://doi.org/10.4161/cc.25380

11. Chen, K., Liu, M.X., Mak, C.S.-L., et al.: Methylation-associated silencing of miR-193a-3p promotes ovarian cancer aggressiveness by targeting GRB7 and MAPK/ERK pathways. Theranostics **8**(2), 423–436 (2018)

12. Delfino, K.R., Rodriguez-Zas, S.L.: Transcription factor-MicroRNA-target gene networks associated with ovarian cancer survival and recurrence. PLoS ONE **8**(3), e58608 (2013). https://doi.org/10.1371/journal.pone.0058608

13. Guo, Y., Sheng, Q., Li, J., et al.: Large scale comparison of gene expression levels by microarrays and RNAseq using TCGA data. PLoS ONE **8**(8), e71462 (2013). https://doi.org/10.1371/journal.pone.0071462

14. Tomczak, K., Czerwińska, P., Wiznerowicz, M.: The Cancer Genome Atlas (TCGA): an immeasurable source of knowledge. Contemp. Oncol. **19**(1A), A68–A77 (2015). https://doi.org/10.5114/wo.2014.47136

15. Troyanskaya, O., Cantor, M., Sherlock, G., et al.: Missing value estimation methods for DNA microarrays. Bioinformatics **17**(6), 520–525 (2001)

16. Xiang, Q., Dai, X., Deng, Y., et al.: Missing value imputation for microarray gene expression data using histone acetylation information. BMC Bioinform. **9**, 252 (2008). https://doi.org/10.1186/1471-2105-9-252

17. Sung, Y.J., Schwander, K., Arnett, D.K., et al.: An empirical comparison of meta-analysis and mega-analysis of individual participant data for identifying gene-environment interactions. Genet. Epidemiol. **38**, 369–378 (2014). https://doi.org/10.1002/gepi.21800

18. Batista, G.E.A.P.A., Monard, M.C.: A Study of K-Nearest Neighbor as an Imputation Method. HIS, University of São Paulo – USP, Brazil (2002)

19. Ahmadi-Nedushan, B.: An optimized instance based learning algorithm for estimation of compressive strength of concrete. Eng. Appl. Artif. Intell. **25**(5), 1073–1081 (2012)

20. Kurum, E., Benayoun, B.A., Malhotra, A., et al.: Computational inference of a genomic pluripotency signature in human and mouse stem cells. Biol. Dir. **11**, 47 (2016). https://doi.org/10.1186/s13062-016-0148-z

21. Chien, C.-H., Sun, Y.-M., Chang, W.-C., et al.: Identifying transcriptional start sites of human microRNAs based on high-throughput sequencing data. Nucleic Acids Res. **39**(21), 9345–9356 (2011). https://doi.org/10.1093/nar/gkr604

22. Griffiths-Jones, S., Grocock, R.J., van Dongen, S., et al.: miRBase: microRNA sequences, targets and gene nomenclature. Nucleic Acids Res. **34**(Database issue), D140–D144 (2006). https://doi.org/10.1093/nar/gkj112

23. Borenstein, M., Hedges, L.V., Higgins, J.P., Rothstein, H.R.: Introduction to Meta-Analysis. Wiley, Chichester (2009)

24. Sung, H.Y., Yang, S.-D., Ju, W., Ahn, J.-H.: Aberrant epigenetic regulation of GABRP associates with aggressive phenotype of ovarian cancer. Exp. Mol. Med. **49**, e335 (2017). https://doi.org/10.1038/emm.2017.62

25. Wang, J., Lu, M., Qiu, C., Cui, Q.: TransmiR: a transcription factor–microRNA regulation database. Nucleic Acids Res. **38**(Database issue), D119–D122 (2010). https://doi.org/10.1093/nar/gkp803

26. Ruepp, A., Kowarsch, A., Schmidl, D., et al.: PhenomiR: a knowledgebase for microRNA expression in diseases and biological processes. Genome Biol. **11**(1), R6 (2010). https://doi.org/10.1186/gb-2010-11-1-r6

27. Jiang, Q., Wang, Y., Hao, Y., et al.: miR2Disease: a manually curated database for microRNA deregulation in human disease. Nucleic Acids Res. **37**(Database issue), D98–D104 (2009). https://doi.org/10.1093/nar/gkn714

28. Ehrlich, M.: DNA hypomethylation in cancer cells. Epigenomics **1**(2), 239–259 (2009). https://doi.org/10.2217/epi.09.33

29. Zhang, W., Barger, C.J., Eng, K.H., et al.: PRAME expression and promoter hypomethylation in epithelial ovarian cancer. Oncotarget **7**(29), 45352–45369 (2016)

30. Armstrong, L.: Epigenetics. Garland Science, New York (2013)

31. Laios, A., O'Toole, S., Flavin, R., et al.: Potential role of miR-9 and miR-223 in recurrent ovarian cancer. Mol. Cancer **7**, 35 (2008). https://doi.org/10.1186/1476-4598-7-35

32. Shell, S., Park, S.-M., Radjabi, A.R., et al.: Let-7 expression defines two differentiation stages of cancer. Proc. Natl. Acad. Sci. U.S.A. **104**(27), 11400–11405 (2007). https://doi.org/10.1073/pnas.0704372104

Author Index

Printed in the United States
By Bookmasters